Praise for

THE
DRAGON
BEHIND
THE GLASS

"With the taut suspense of a spy novel, Voigt paints a vivid world of murder, black-market deals, and habitat destruction surrounding a fish that's considered, ironically, to be a good-luck charm."

—*Discover*

"Voigt relates her continent-hopping adventures as she struggles to make sense of 'a modern paradox: the mass-produced endangered species.' . . . What follows is an immensely satisfying story, full of surprises and suspense."

—*The Wall Street Journal*

"Many a true-crime study could be attributed to an author's honest enthusiasm for weirdness. (I'm thinking of *The Orchid Thief*, Susan Orlean's wondrous, strange book about an orchid poacher's bizarre search for the rare ghost orchid that grows in the swamplands of Florida's Fakahatchee Strand State Preserve.) *The Dragon Behind the Glass* . . . is the same kind of curiously edifying book. . . . Comprehensively researched and gracefully written."

—Marilyn Stasio, *The New York Times Book Review*

"This book starts with a 'bang'—a murder to be exact—and the momentum just keeps going from there. *The Dragon Behind the Glass* is a gripping blend of investigative journalism, science, international crime, travelogue, and history. . . . You don't need to know anything about tropical fishes or fishkeeping to be totally riveted by this informative page-turner."

—*Forbes*, The 10 Best Conservation and Environment Books of 2016

"[An] engaging tale of obsession and perseverance . . . Voigt's passion in pursuing her subject is infectious, as is the self-deprecating humor she injects into her enthralling look at the intersection of science, commercialism, and conservation."

—*Publishers Weekly* (starred review)

"Voigt's passionate narrative perfectly conveys the obsessive world in which [the arowana] swims."

—*Publishers Weekly*, Best Summer Books of 2016

"Not since Candice Millard published *The River of Doubt* has the world of the Amazon, Borneo, Myanmar, and other exotic locations been so colorfully portrayed. . . . A compelling must-read."

—*Library Journal* (starred review)

"Voigt immerses readers in the ruthless world of the exotic fish trade that involves murder, smugglers, and globe-trotting. More than just a compelling tale, though, this is a perceptive work that examines layered themes about endangered species and our relationship to nature."

—*Library Journal*, Best Summer Books of 2016

"A spirited debut . . . A fresh, lively look at an obsessive desire to own a piece of the wild."

—*Kirkus Reviews*

"Who would've thought the history of a rare fish could be so enthralling? Voigt traces the bizarre story of the world's most expensive aquarium fish, the Asian 'dragon fish,' in a story that reads more like fiction, what with all the murder, smuggling, and general intrigue."

—*PureWow*, The Ultimate 2016 Summer Book Guide

"A masterpiece! Emily Voigt has raised the bar for anyone who thinks they can tell a good fish story. What an extraordinary and extraordinarily well-told tale. Voigt brings such wonderful humor, adventure, and hard science to this subject, I found myself unable to put the book down. Never has science been so much criminally good fun. I will never look upon a goldfish the same way again."

—Bryan Christy, author of *The Lizard King: The True Crimes and Passions of the World's Greatest Reptile Smugglers*

"Few writers can match the intelligence, charm, wit, and sheer audacity that Emily Voigt brings to bear in this highly readable and important book. From the bleak housing projects of the South Bronx to the steamy jungles of southern Myanmar, Voigt takes us along on a journey of adventure and discovery in her quest to find an increasingly rare fish in the wild. With a page-turning plot and a cast of vivid characters, *The Dragon Behind the Glass* shines a powerful light on the international trade in endangered species."

—Scott Wallace, author of *The Unconquered: In Search of the Amazon's Last Uncontacted Tribes*

Illustration accompanying the species description of the
Asian arowana by Salomon Müller and Hermann Schlegel, 1840.

THE
DRAGON
BEHIND
THE GLASS

A TRUE STORY OF POWER,
OBSESSION, AND THE WORLD'S
MOST COVETED FISH

EMILY VOIGT

SCRIBNER
New York   London   Toronto   Sydney   New Delhi

SCRIBNER

An Imprint of Simon & Schuster, Inc.
1230 Avenue of the Americas
New York, NY 10020

First Scribner trade paperback edition May 2017

SCRIBNER and design are registered trademarks of
The Gale Group, Inc., used under license by Simon & Schuster, Inc.,
the publisher of this work.

For information about special discounts for bulk purchases,
please contact Simon & Schuster Special Sales at 1-866-506-1949
or business@simonandschuster.com.

The Simon & Schuster Speakers Bureau can bring authors to
your live event. For more information or to book an event, contact
the Simon & Schuster Speakers Bureau at 1-866-248-3049
or visit our website at www.simonspeakers.com.

Interior design by Kyle Kabel

Manufactured in the United States of America

3   5   7   9   10   8   6   4

The Library of Congress has cataloged the hardcover edition as follows:

Names: Voigt, Emily, author.
Title: The dragon behind the glass : a true story of power, obsession, and
the world / Emily Voigt.
Description: First Scribner hardcover edition. | New York, NY : Scribner,
2016. | Includes bibliographical references and index.
Identifiers: LCCN 2015040075| ISBN 9781451678949 (alk. paper) | ISBN
9781451678963 (alk. paper) | ISBN 9781451678956 (alk. paper)
Subjects: LCSH: Scleropages formosus.
Classification: LCC QL638.O88 V65 2016 | DDC 597.176—dc23
LC record available at https://lccn.loc.gov/2015040075

ISBN 978-1-4516-7894-9
ISBN 978-1-4516-7895-6 (pbk)
ISBN 978-1-4516-7896-3 (ebook)

ART CREDITS

Title page: Courtesy of The Naturalis Museum, Leiden, Holland;
p. 283: Courtesy of The Natural History Museum, London

For Jeff
&
For my mother and father

The truth is that we never conquered the world, never understood it; we only think we have control. We do not even know why we respond a certain way to other organisms, and need them in diverse ways, so deeply.

—Edward O. Wilson, *Biophilia*

Contents

CONTENTS

A Note on Names

This story spans fifteen countries across which naming practices vary widely. In traditional Chinese names, for example, the family name precedes the given name, while many Indonesians have only one name. When deciding how to refer to someone—whether by first or last name where the choice exists—I have done so based on what I feel will be most memorable to the reader while aiming for consistency and taking into account my relationship to the person.

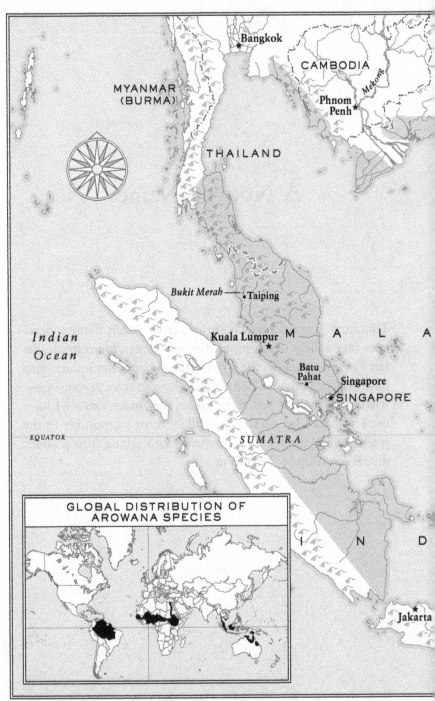

Sources: Ng Heok Hee, Maurice Kottelat, and Tim M. Berra

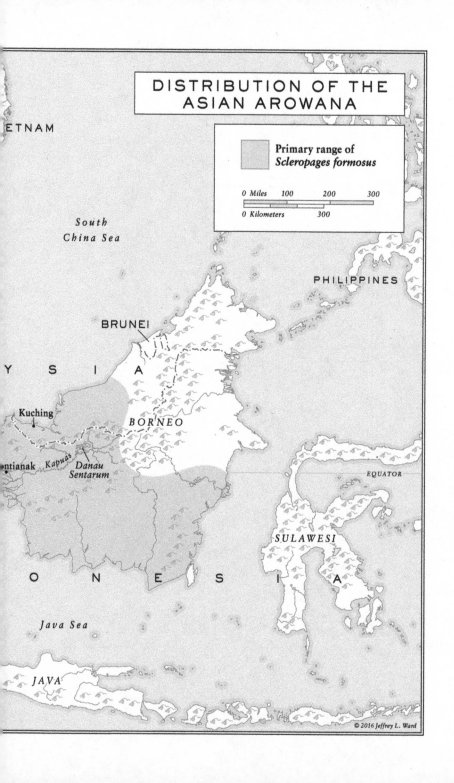

DISTRIBUTION OF THE ASIAN AROWANA

Primary range of
Scleropages formosus

0 Miles   100   200   300

0 Kilometers   300

ETNAM

South
China Sea

PHILIPPINES

BRUNEI

Y   S   I   A

Kuching

BORNEO

ntianak   *Kapuas*   Danau
Sentarum

EQUATOR

O   N   E   S   I   A

SULAWESI

Java Sea

JAVA

© 2016 Jeffrey L. Ward

Bersatu, he saw that Kuan Aquarium was still shuttered. But as he slipped off his motorbike and moved closer to the door, he noticed the padlock was missing. Lifting the grille, he stepped gingerly into the dark, narrow space crowded with gurgling aquariums that cast a dim fluorescent glow. When his eyes adjusted to the light, he saw that the shop was tidy as ever, goldfish and guppies swimming idly in their tanks.

Taking a step forward, however, his shoe crunched on broken glass, and he looked down to see a streak of crimson on the concrete floor. Blood led to more blood, tracks suggesting a wounded animal trying to escape on all fours. What the father encountered at the end of this grisly trail would sear itself in his mind, waking him in the night for the rest of his life: his sweet boy lying crumpled on the ground, stabbed ten times. He was dead—viciously slashed, his neck slit so deeply he was nearly decapitated.

The old man began to shake uncontrollably as he stumbled outside, calling for help. But it was too late. The police would find neither a murder weapon nor a single fingerprint. The victim's wallet was still in his pocket, contents intact. Only one thing was missing. Above Chan Kok Kuan's lifeless body, the tanks of the Asian arowana loomed empty. All twentysome dragon fish were gone.

PART I

―――――

"DRAGONS IN THEIR PLEASANT PALACES"

(ISAIAH 13:22)

CHAPTER ONE

The Pet Detective

NEW YORK

On a freezing Tuesday in March 2009, my alarm blared at 4:00 a.m. By 6:45, I stood shivering outside a housing project in the South Bronx with Lieutenant John Fitzpatrick and three junior officers, fresh-faced graduates of the Academy. The entire scene was gray—the potholed roads, the sooty snow, the late-winter sky—except the officers themselves, who provided the only glimmer of green. Rather than standard NYPD issue, they wore olive uniforms and trooper hats, à la Ranger Smith from *The Yogi Bear Show*. As they crunched across the unshoveled walkway, a passing teenage girl wisecracked, "Ain't you supposed to be in the forest?"

Fitzpatrick, who had been patrolling the same beat since 1996, ignored her, keeping his eyes trained on one of the brick high-rises lined up like dominoes. As a cop (of sorts) from Brooklyn, descended from a clan of cops from Brooklyn, he looked the part, a towering man of forty-one with a crew cut and dimpled chin. Tucked under his arm was a file containing a photograph of the suspect he was after—someone he believed could be armed and dangerous.

Inside, the building's lobby was dimly lit and gloomy. The elevator clattered open, and we crowded in, squeaking up to the eighth floor, where the officers' boots echoed down the long hall before halting outside one apartment. Fitzpatrick pounded on the door. After half a minute passed and nothing happened, he raised his fist again, pounding harder and longer. A baby cried down the hall. At last, a male voice, gravelly with sleep, croaked, "Who is it?"

"State Environmental Police," Fitzpatrick announced.

"*Who?*" said the voice, sounding genuinely confused.

The door cracked open to reveal a stocky young man with full-sleeve tattoos wearing flannel pajama bottoms, his eyes squinting against the light. His name was Jason Cruz. Asked if he knew what brought the officers to his door, he shook his head no and said, "I don't at all."

"We're here," Fitzpatrick enlightened him, "because of the alligator that you were offering to sell on Craigslist."

I WAS REPORTING a story about exotic pets for a science program on NPR, and it had taken me six months to get permission to join Fitzpatrick, a detective with the New York State Department of Environmental Conservation, on one of his busts. When I'd first called him the previous summer, I'd found him brimming with bizarre tales from an urban bestiary. During his time policing the city's illegal wildlife trade, he had encountered everything from gorilla-hand ashtrays to twelve hundred turtles crammed into a swank Tribeca loft, their owner left with no room for a bed. There was the Harlem man who kept Ming the tiger and Al the alligator in the apartment where his mother was raising eight small children; the wealthy Brooklyn family who treated their African Diana monkey, one of the rarest primate species on earth, like a second daughter—even threatening to barricade their home if marshals tried to take her away; and the proprietor of a popular curio store in SoHo who landed in prison for selling not only a chimpanzee skeleton and walrus tusks but also human body parts. From time to time, a notorious dealer called Mr. Anything Man surfaced in his jalopy advertising exotic live animals, dead or alive.

Anything. It's the theme of the city's wildlife trade. Sure enough, when Fitzpatrick asked the pajama-clad Cruz how he came by the alligator he'd been trying to sell, Cruz shrugged and said, "This is the Bronx. You can get anything."

Inside his apartment, which was small and tidy with a black leather couch, flatscreen TV, and mirrored photograph of the Manhattan skyline, a dog barked from the bathroom. Birds tweeted in the kitchen. Fitzpatrick walked over to a trio of tanks and inspected a leopard gecko in one and, in the other, a teaspoon-size baby Nile monitor—a yellow-striped lizard with a blue, forked tongue fond of devouring cats when full grown. The third tank was empty, except for a few inches of stagnant water, which Fitzpatrick leaned in to sniff suspiciously. "So where is the alligator now?" he asked.

"I gave it back," Cruz said, claiming he'd bought the reptile from a stranger outside PetSmart on Pelham Parkway. Though it was only a foot long, his girlfriend had pointed out its likelihood of enlarging and insisted that it had to go for the sake of their two-year-old daughter. After this ultimatum, Cruz said he tried to unload the gator online, but Craigslist flagged the ad; so he returned it to the dealer who sold it to him in the first place. "He was a Puerto Rican dude," he offered.

Fitzpatrick jotted this on a notepad. "Now it's down to just a few million people in the Bronx," he said drily.

I was as disappointed as Fitzpatrick to have narrowly missed the alligator. I'd been hoping for a scene like the time he taped up the snout of a three-foot-long caiman and drove it thrashing in his front passenger seat to the Bronx Zoo. What's more, Cruz didn't live up to Fitzpatrick's billing of the typical alligator aficionado as an exemplar of machismo and aggression. Pet alligators were supposed to be particularly hot among gang members and drug dealers, but Cruz didn't seem like either. Before his daughter was born, he used to keep pit bulls, as evidenced by black leather harnesses with metal studs hanging from the wall; but the dog barking in the bathroom turned out to be a poodle.

"You can really get in trouble over, like, an alligator?" Cruz asked Fitzpatrick, still bewildered by what was happening.

His pregnant girlfriend, who had emerged from the bedroom, yawning and looking unamused, added, "There's a lot of people that sell alligators."

"It's a criminal offense," Fitzpatrick told them, explaining that New York State prohibits the commercial sale of live crocodilians, while the city goes further, banning just about every exotic pet from scorpions to ferrets to polar bears. The Nile monitor was illegal too and would have to be seized. Cruz looked crestfallen as his girlfriend found an empty shoebox, into which he gingerly lifted the tiny lizard, wrapping it in a T-shirt to protect it from the cold.

"There's not many animals you can keep here," Fitzpatrick advised him, "except a dog, cat, goldfish, canary—"

"I got like twenty lovebirds!" Cruz exclaimed.

In the kitchen, Fitzpatrick inspected the stacked cages of small, green parrots with yellow chests and red beaks, deeming them permissible.

"If I didn't have kids or nothing, I would've had cobras here, vipers, all types of stuff," Cruz said wistfully, explaining that he had loved animals since he was a child, particularly after escaping the Bronx to visit his aunt in Florida where alligators sunned themselves in the backyard and took dips in the swimming pool. While waiting for the officers to write up a court summons, he called his mother and told her that he was going to be on Animal Planet.

"NPR," I mouthed, then frowned, recalling how my producer had requested high drama—something along the lines of a wildebeest in Queens.

After hanging up, Cruz turned to me and grew philosophical: "You know what it is? You like animals, and you get tired of seeing the same animals over and over. You go to the pet store, and they have this and that—and you know everybody's got it. So you try to get something different."

"Honestly, that's part of the problem," Fitzpatrick said as he handed over the summons. "Then you get into endangered species."

But Cruz, marveling at how easily he could acquire a fifteen-foot anaconda rumored to be for sale, didn't seem to hear.

MORE EXOTIC ANIMALS are believed to live in American homes than in American zoos. Yet the desire of someone like Cruz to keep an alligator in his living room defies classic theories of pet-keeping, which hold that humans keep pets for unconditional love, for example, or because a misdirected *cute response* (the scientific term) compels us to care for other species the way we do our own offspring. Alligators are neither affectionate nor cute, at least not in the sense of being cuddly and having large eyes, a round face, and an oversize head like a human infant or a pug dog. Rather, the appeal seems to lie in the opposite direction—the alligator's ferocity, its wild and untamable nature.

In a way, what Cruz had assembled in his Bronx apartment could be seen as having a long historical precedent. It was a menagerie, a collection of exotic creatures kept in captivity for exhibition. Menageries first appeared with the advent of urbanization, when contact with wild animals became rare, and the keeping of exotics was almost exclusively the privilege of royalty and nobility. Mesopotamian kings, who received foreign beasts as tribute, created elaborate gardens to house them called *paradeisoi*, later to serve as the model for the biblical Garden of Eden. Egyptian pharaohs collected baboons, hippos, and elephants from sub-Saharan Africa and had them mummified to take into the afterlife. In the classical world, the Greeks brought back wild animals from military expeditions, a tradition their Roman conquerors continued, slaughtering the beasts in public arenas, a popular entertainment for nearly five hundred years. The Tower of London gained its famed royal menagerie in 1204. Across the Atlantic, Lord Moctezuma dazzled the first conquistadores with his magnificent pleasure gardens, replete with rare aquatic birds and wildcats tended by their own physicians. With the European discovery of the New World, the desire to own exotic animals intensified, and the Renaissance saw the invention of the private "cabinet of curiosities." By the eighteenth century, the aristocracy was clamoring for monkeys and parrots as novel playthings.

The human species is unique in its compulsion to tame and nurture nearly all other vertebrate animals. In his 1984 classic, *Dominance & Affection: The Making of Pets*, the cultural geographer Yi-Fu Tuan characterizes this proclivity as an exercise in power—a kind of playful domination stemming from our desire to control the unpredictable forces of nature. According to Tuan, the keeping of pets reflects our hunger for status symbols, for what the philosopher Jean-Paul Sartre called the "carnal, clinging, humble, organic, milky taste of the creature," which underlies all luxury goods.

The modern pet shop first appeared in American cities in the 1890s; and with it began the mass importation of exotic animals from Asia and South America. Pet-keeping in the United States exploded in the economic boom following World War II and, since the 1970s, has more than tripled. For the first time in history, more American households have pets than don't, including some 86 million cats and 78 million dogs. But no one knows how many "exotics" there are, not least because no one agrees how to define an exotic pet. Does a native but wild animal like a skunk count? What about a potbellied pig? Further complicating matters, much of the trade operates underground, flouting state and federal laws.

In the 1980s, wild birds comprised the hottest segment of the black market. Next to take off was the "herp trade," short for herpetoculture, which is the keeping of reptiles and amphibians such as turtles, lizards, and salamanders. "That's where you have the big-buck items," Fitzpatrick told me. He described how he once went undercover posing as a reptile collector to buy a critically endangered $18,000 radiated tortoise from Madagascar with a brilliant star pattern on its shell. As he was making the purchase, the dealer showed him a picture of a one-of-a-kind turtle—an albino river cooter, its entire body a pale jade white—and said he'd been offered $101,000 for the animal but was holding out for more.

On that occasion, wary of being searched, Fitzpatrick had slipped his gun into a drawer in another room, and when his backup team was delayed, he began to sweat. Though the dealer didn't resist arrest, high-stakes turtle trafficking can be tied up with all sorts of unsavory behavior. Interpol warns that organized-crime networks use the same routes to smuggle animals as they do weapons, drugs, and people—that environmental crime goes hand in hand with corruption, money laundering, even murder.

IT WAS FITZPATRICK who first told me about the Asian arowana. At the time, I still thought of pet fish as one step up from potted plants. Had someone informed me that fish comprise the vast majority of exotic pets—that they are the most common pet, period, with more than 100 million swimming in aquariums across the United States—I would not have cared one bit. If there was anything appealing about them, it was their comic irrelevancy and their association with childhood.

When I was little, my parents got a tank of goldfish as a sorry substitute for a dog or cat. One day when I was about six, I noticed bubbles escaping from the fish's puckering mouths and wondered if they were talking to each other. Retrieving my Fisher-Price stethoscope, I pushed a chair next to the aquarium and climbed up to listen for tiny voices rising from the water's surface. All I heard was silence. After that, I ignored them, and they continued to ignore me.

Not everyone, however, shared my dispassion. "One thing we deal with here in the city is a fish," Fitzpatrick told me. "Arowana." I misheard this as "marijuana," and the association proved surprisingly apt. Protected by the Endangered Species Act, the Asian arowana cannot legally be brought into the United States as a pet. Yet trafficking is rampant across the country. Fitzpatrick recounted a bust at a dingy Brooklyn sweatshop, where women sat hunched over sewing machines, scraps of fabric strewn about the floor—a front for running fish. Another time, acting on a tip, he caught a Malaysian-born Queens man smuggling arowana through JFK Airport in water-filled baggies packed in a suitcase. Fitzpatrick noted that even the cheapest specimens sold for thousands of dollars, and prices went up from there, depending on coloration, the most desirable being red. "In certain Asian populations the arowana is considered good luck or a sign of prosperity or a status symbol," he explained. "And it's something that's been overharvested for the pet trade for those reasons."

The obsession with the fish, however, wasn't limited to Asian cultures. There was, for example, the Wall Street banker who broke down crying after he was arrested for possession of the species, confessing he couldn't resist its dark-alley appeal. "In recent years, we've seen more cases involving non-Asians—white people," Fitzpatrick said. The previous summer, two Long Island men had been caught at the Canadian border, driving back from Montreal with four specimens swimming in the spare-tire well of their SUV. Then a young man was busted running arowana from his family's home in the suburbs. Fitzpatrick theorized that selling black-market fish was considered safer than dealing in other high-value contraband such as drugs and guns—especially in New York, where just twenty environmental conservation officers cover the same territory as some thirty-four thousand NYPD.

The way he saw it, wildlife traffickers were motivated by pure greed, participating in what may well be the world's most profitable form of illegal trade. But the collectors were driven by a passion he found easier to relate to. "I think that a lot of the people who have these animals are interested in nature, and that, in and of itself, is not a bad thing," he said. "It's just they're going about it the wrong way."

As a young graduate student in biology, Fitzpatrick had spent two months studying birds in the jungles of Venezuela, where he lost thirty pounds, grew "a full Grizzly Adams beard," and got all sorts of weird skin infections before realizing he was really a city person. He still loved tropical animals—from afar. His only pet was a pint-size Maltese with a name he refused to disclose. "Like Snowball?" I asked.

"Something along those lines," he said.

"Flufferbutter?"

"You get the picture."

THAT EVENING AFTER I got home, I looked up the Asian arowana to see what more I could find. Fitzpatrick had mentioned that the species was officially called the Asian bony-tongue—"very unsexy name"—for a long bone of a tongue, bristling with prickly, pinlike teeth, which the fish uses to seize and crush prey against teeth on the roof of its mouth.

The bonytongues, I learned, are among the most ancient living fish on earth. The oldest fossils date to the Late Jurassic or Early Cretaceous and reveal giant creatures with ferocious fangs that roamed the prehistoric seas, preying on ichthyosaurs and plesiosaurs. Today the family Osteoglossidae (from the Greek *osteo* meaning "bone," and *gloss* meaning "tongue") no longer inhabits the oceans but rather the rivers and lakes of the earth's tropical midsection, where it traditionally includes one of the largest of all freshwater fish: the Amazonian arapaima, which can grow nearly fifteen feet long and weigh some 450 pounds. With the exception of this giant, the rest of the family—all long, thin creatures armored in a mosaic of large, heavy scales—are commonly known as arowanas.

The most formidable among them (or at least the most acrobatic) is the South American silver arowana, also known as the water monkey for its ability to leap six feet into the air to snatch bugs, birds, snakes, and bats from overhanging branches. (Do not google *arowana eats duckling*.) In 2008, when a New Jersey man reached into a tank at Camden's Adventure Aquarium to touch a silver arowana, the fish tried to make a meal of his hand. In his subsequent lawsuit, the victim alleged "painful bodily injuries" and that his three-year-old son suffered "severe emotional distress, headaches, nausea, long continued mental disturbance and repeated hysterical attacks" after witnessing the incident.

Despite or perhaps because of its ferocity, the silver arowana is a popular pet in the United States—and a perfectly legal one—with young fry selling for as little as $30 to $50 apiece. In all, there are seven recognized arowana species (with three more disputed among scientists) in South America, Africa, Southeast Asia, New Guinea, and Australia. Only the Asian variety is banned in the United States.

Overseas, however, the Asian arowana is an openly coveted commodity in a legitimate luxury market. "Forget oil and diamonds, the next big thing in Southeast Asia is fish," I read in the hobbyist magazine *Practical Fishkeeping*, which described how fifty specimens collectively valued at a million dollars had been placed under twenty-four-hour guard in Jakarta, Indonesia. "While these fish may be disappearing in the wild, their popularity amongst Asia's richest is ever increasing."

In some instances, the species was reared on farms that could pass as prisons—facilities protected by nested walls, watchtowers, and rottweilers that prowled the perimeters at night. The reason for the heavy security became clear as I dug deeper into an international news archive that chronicled a spate of fish thefts across Southeast Asia. In Malaysia, five arowana stolen from a woman's house were worth more than all her other possessions combined. Singapore, which boasts one of the lowest crime rates in the world, once reported four such heists in a single week. One thief punched out an elderly woman who chased him as he made off with her prized fish in a sloshing bucket. He was sentenced to three years in prison and twelve strokes of a wet cane.

As for who was taking the fish, some surmised that the thieves were fish lovers who could not afford the astronomical prices. Others suspected that a crime syndicate was behind the thefts—a sort of shadowy "fish mafia." Bolstering this theory was a harrowing case in Indonesia, where an arowana dealer and his heavily armed cronies allegedly kidnapped and held for ransom a Japanese buyer.

Despite all this criminality, however, the trade in the farmed fish was legal not only in Asia but throughout most of the world, including Canada and Europe—the one major exception (other than the United States) being Australia, which bans the species to protect its own tropical fauna. A few years back, a forty-five-year-old housewife was arrested at the Melbourne Airport trying to enter the country with fifty-one fish, including an Asian arowana, hidden beneath her poufy skirt. "We became suspicious after hearing these flipping and flapping noises," a customs official later told the press.

Such absurd tales of smuggling kept me up late as I pored over the hundred-some articles I'd printed out and spread across my living room floor. The picture that gradually emerged, however, looked less like the illegal drug trade and more like a parody of Manhattan's overheated art scene, complete with record-breaking prices, anonymous buyers, stolen specimens, unsavory dealers, and even clever fakes. Whatever the best metaphor, it seemed the Asian arowana had a long history of driving human beings to dangerous extremes.

One summer in college, I'd read Jane Goodall's *In the Shadow of Man*, and ever since, I'd dreamed of venturing into the jungle to write a great story about wildlife. Around the time my curiosity in the arowana began, I'd been awarded a fellowship intended to fund a reporting project abroad. Now I knew how I'd use it: I would go see for myself where all these smuggled arowana were coming from and what made the species so irresistible. Goodall had her noble, tool-wielding primates— I would have a bad-tempered, bony-tongued fish.

It was obvious where to start. At the center of the glamorous world of Southeast Asian aquaculture reigned a flamboyant Singaporean tastemaker known as Kenny the Fish, a chain-smoking millionaire fond of posing nude behind strategically placed aquatic pets. The Fish's real name was Kenny Yap, and he was the executive chairman of an ornamental-fish farm so lucrative that it was listed on Singapore's main stock exchange. Recently, the Singaporean press had dubbed him one of the city's most eligible bachelors and called for him to host a national spin-off of Donald Trump's reality show, *The Apprentice*. To enter his website, I had to click on his belly button.

I've since come to think of this navel as the rabbit hole into which I fell, not to emerge for some three and a half years.

CHAPTER TWO

The Fish

"Welcome to our Island Nation!!!" Kenny e-mailed me the Saturday I arrived in Singapore, the southernmost tip of the Asian mainland, a dot on the exclamation point of the Malay Peninsula. "See you on Monday and do have a fun and naughty weekend! The Fish."

It was May, two months after the failed alligator bust, and I could scarcely have been—or felt—farther from the South Bronx. Singapore is an almost eerily clean and sanitized city, largely traversable through air-conditioned tunnels connecting a vast maze of malls. That weekend, I wandered through the gleaming shopping district down to the waterfront, where I came across a giant white statue with the body of a fish and the head of a lion, spewing water into Marina Bay. It was the Merlion, Singapore's national symbol, dreamed up by the tourist board in the 1960s. In 1979, native-born poet Edwin Thumboo imagined Ulysses encountering the trademarked logo and declaiming, "Nothing, nothing in my days / Foreshadowed this / Half-beast, half-fish."

Singapore's *other* half-fish chimera—Kenny the Fish—seemed

to be nearly as famous around town as the Merlion. Come Monday, my cabdriver was not happy to find himself heading to a remote agrotechnology park on the northwest side of the island—until he learned I was going there to meet Kenny. "I saw him once at a coffee shop," the driver told me excitedly. "He's grown his family business into a big success."

Kenny's ornamental-fish company, Qian Hu Corporation Ltd. (pronounced *chien who*, Chinese for "a thousand lakes"), was located off a winding road, its leafy entrance marked by a statue of a lounging blue-and-white suckerfish with a sweet, dopey expression. At the base of the drive sat warehouses with wet concrete floors and an outdoor café where Chinese pop songs blared through loudspeakers. A sign pointed to a fish spa with pools of minnows that nibbled dead skin off visitors' toes, and wading ponds where children could net small fish. Painted powder blue, the entire facility had a Disneyesque feel. Practically every available surface was plastered with newspaper clippings featuring Kenny wearing the same practiced expression of ecstatic delight.

I found the executive chairman himself in the corner office of a modest, one-story building, where he sat behind a hot-pink-and-turquoise desk. At forty-four, he was as skinny as a teenager, with a round, boyish face that lit up when he saw me. He wore jeans and a white short-sleeved shirt emblazoned with the suckerfish I'd seen atop the drive. Five photographs hung in a row above his head. They appeared to be of Kenny, posing nude with fish.

"*Half*-naked," he corrected me, pointing to a shot in which he was seated spread-eagle, his glistening arms embracing a strategically placed bag of koi. "Is that nude?"

"I can't tell," I said honestly. "It could be."

"Okay. No, I did wear underwear, and then I just have a bag of fish there." Kenny explained he'd been getting a passport photo taken when he got the idea to spice up the photo shoot and strip down to a G-string. "I'm naughty, I'm playful, you know?" he said, then offered me a bottle of water and leaned back in his chair, twisting one open himself. "I need water. Last night I had a lot of drinks."

It certainly wasn't by accident that Kenny became a cultural icon. A marketer par excellence, he made up his own nickname and began publishing an annual calendar featuring his own quotes to promote his fish.* Yet his playboy image, however carefully crafted, owed much of its appeal to his rise from humble origins as the youngest of nine children born to a pig breeder. When the cleanliness-obsessed Singaporean government phased out pig farming in the early eighties, his family converted their pens into concrete ponds and started breeding guppies. In 1989, while Kenny was abroad studying marketing at Ohio State University—the first in his family to attend college—one of his brothers called to tell him that heavy rains had washed away all the guppy ponds. The following year, he returned and joined in rebuilding the family farm.

The brothers invested all their resources in a batch of four thousand Chinese high-fin banded loaches, a bottom-feeder from the Yangtze River that was particularly hot in the aquarium trade at the time (though supremely unsuited for home captivity, since it grows to the size of an alligator; it landed on Lieutenant Fitzpatrick's Most Wanted list). But all the fish died in two to three weeks, apparently in reaction to vibrations from construction on the farm. After that, Kenny decided the species would serve as the company's logo—the friendly suckerfish at the head of the drive and on his shirt—providing a reminder of the importance of diversifying one's aquatic portfolio.

"It's not about relying on one fish," Kenny told me, pointing out that his company now exported more than a thousand

---

* An example: "Not practical thought such as hoping to be invisible so that you can anyhow touch people is like climbing a tree to catch a fish."

species to eighty different countries. Still, the Asian arowana accounted for more than 20 percent of fish profits, a fact that Kenny attributed to its high value. He said he didn't know anything about the arowana being smuggled into the United States, which he dismissed as a "cheapskate market."

"What's the most expensive arowana you've ever sold?" I asked.

"Seventy thousand dollars US for a big one," Kenny said nonchalantly, noting that a Japanese buyer had purchased three such specimens many years ago, a trio of "good-quality red arowana," aged seven or eight. That's young adulthood for the fish, which can live many decades, though they often die prematurely in tanks.

"To be fair, not all cost that much," Kenny added, explaining that arowana usually sell around the age of six months, when they're roughly the length of a pencil and typically fetch about $1,000 to $2,000. "People want to rear them from small to cultivate a certain kind of relationship." In the past, he'd told the press that an arowana could be trained like a dog or cat to "stay by the owner's side when he is unhappy" (never mind that it's confined to a tank). The flip side to this intimacy was that the fish was prone to temper tantrums and could behave "like a spoilt child."

Certainly, Kenny was aware that most aquatic fads were fleeting. In 2001, for example, the Asian fish world went crazy over a new hybrid called the *louhan* or flowerhorn, which looked like a tilapia with a bulbous blossom of a pompadour, as if the fish were wearing its brain outside its head. Somehow a rumor started that the unique markings on the flanks of this creature predicted winning lotto numbers. Soon flowerhorns began to fetch exorbitant sums, the best specimens commanding tens of thousands of dollars. At the height of the craze, the number of fish shops in Singapore more than tripled in a single year. There was a rush on tanks. Fish food was scarce. And then, abruptly, the bubble burst around 2005. Tanks littered the streets. Shops went under. It was an episode right out of Charles Mackay's 1841 classic *Extraordinary Popular Delusions and the Madness of Crowds*.

Kenny, however, insisted the fate of the flowerhorn could never befall the Asian arowana, which had far greater staying power for two reasons. The first was its evocative name: the dragon fish. In Chinese culture, the mythical creature is not a fearsome monster but a beloved and beneficent symbol. "The Eastern dragon is considered auspicious, prosperous, good, unlike the Western one," Kenny explained, leaning forward and placing his palms flat on his desk. "The Western dragon is always *evil*, blowing *fire*, *kill* people—*no!*" He slammed his fist down, then erupted in laughter. "Secondly"—he resumed his composure—"this fish is very rare. You can't really flood the market."

JUST HOW RARE, however, was not entirely clear. In the scientific world, the Asian arowana had long been considered one species, *Scleropages formosus*, with a wide distribution in rivers and lakes throughout Southeast Asia. Its natural range was believed to extend from southern Vietnam, Cambodia, Thailand, and possibly Myanmar, down the Malay Peninsula to the islands of Sumatra and Borneo.

Complicating matters, however, dragon fish come in a riot of color varieties, often with overlapping names—blood red, chili red, golden crossback, yellowtail silver, graytail silver, blue Malayan, redtail golden, and more. Some are much rarer and considered more desirable than others. The most highly coveted—or at least the traditional favorite—is the legendary Super Red, native to a single remote lake system in the heart of Borneo. By contrast, the green, which doesn't merit a fancy ornamental name, is the most common. As late as 2008, researchers with Conservation International reported that rural Cambodians were still eating greens, even as wild populations plummeted due to overharvesting for the aquarium trade.

Actually, not that long ago *all* varieties of the Asian arowana were more often eaten than kept as pet fish. In 1964, when the International Union for Conservation of Nature (IUCN) set up a special group dedicated to the survival of freshwater fishes, the Asian arowana was still primarily "an important source of food to people living in the swamp," and a popular game fish among anglers in Malaysia. Due to reports of its declining numbers, however, it was added to the IUCN Red Data Book of threatened species. The listing led to its inclusion on the very first Appendix I of the Convention on International Trade in Endangered Species of Wild Fauna and Flora (CITES). Today CITES (pronounced *sigh-tees*) is among the largest conservation agreements in the world, regulating the movement of endangered species across the borders of 186 member states. The animals and plants on Appendix I are considered the rarest of the rare and generally banned from international trade. In 1975, when the treaty took effect, such was the case for the Asian arowana.

The ban didn't make a lot of sense because the fish *wasn't* being traded internationally at that time, at least not on any large scale. Locally, however, aquarists had begun to notice it. In 1967, Chew Thean Yang, the proprietor of an aquarium shop in northern Malaysia, was driving through the jungle to collect small fishes when he stopped at a roadside market and spotted a golden arowana for sale. The fish was dead; but he was struck by its beauty, admiring its shiny scales and the pair of barbels jutting from its chin. He asked the fisherman to catch him a live one, which he displayed in his shop on the island of Penang. Calling the creature *jīnlóng yú*, the golden dragon fish, Chew was soon selling juveniles for six Malaysian ringgit apiece (about $2.00 US).

The golden and green varieties were the only ones known to the aquarium trade until the late 1970s, when Indonesian loggers cutting deep into the virgin forests of Borneo's interior discovered the Super Red. In the early 1980s, smugglers snuck the fish out of Indonesia to Taiwan, where it became the object of a virtual cult. Not only is red a favorite color in Chinese culture, but fish themselves are considered auspicious according to the principles of feng shui. As Kenny explained to me, the Chinese word for "fish," *yú,* is pronounced the same as the word for "surplus"; and "water," *shuǐ,* is slang for "wealth" in Cantonese.

Yet when I asked Kenny if he believed the arowana was lucky, he chose his words carefully: "Keeping fish gives you a certain kind of tranquillity. I think if you are happier, if you're more relaxed, you might be able to make better decisions. Your family relationships might be better. So the whole process might bring you better luck."

"But nothing about the arowana specifically?" I pressed.

"Oh, there are stories," he said. "Because this fish—I think they attach to their owners. They sense some things that we still do not know. . . ."

Since arowana are naturally great jumpers that rocket out of the water to catch prey, they often leap from uncovered aquariums and are later found dead on the floor. Perhaps to rationalize the loss of such a substantial investment, hobbyists have come to imbue this behavior with a Christlike aura, suggesting the fish gives its life to save its owner. Several such anecdotes were posted around Kenny's farm. One, called "The Lucky Arowana," related how the fish of a rich Indonesian businessman vaulted to its death to prevent him from loaning a large sum of money to an untrustworthy associate (the businessman correctly interpreted his dead fish as a bad omen and reneged on the loan). This "sacrificial suicide" is the most frequently recurring motif in stories about the fish. But there are many others as well: the arowana that protects an apartment from fire or burglars; the arowana that breaks up a marriage by harassing the wife whenever the husband is out; even the arowana that kills—one man was said to have been found dead in a canal after skimping on fish food.

IN THE MID-1980s, the black market boomed and the cult of the dragon fish spread from Taiwan to Japan, where the Asian arowana was supposedly smuggled in whiskey bottles. Meanwhile, with wild populations plummeting, farmers throughout Southeast Asia had begun breeding the fish—a feat previously believed impossible following a failed attempt at culturing the species for food in the 1920s. Now skyrocketing prices motivated farmers to figure out how to make it work: dig ponds directly into the earth, fill them with dark, tannic water, and remove any small fry before the older fish eat them. In 1990, CITES moved all Indonesian populations of the Asian arowana to the less restrictive Appendix II, which allowed farmed specimens to be sold across international borders. The idea was to kill the illicit trade with a legitimate one.

But it didn't work. The poaching of wild fish continued, and in 1995 all populations went back on Appendix I. By that time, however, CITES had agreed to try an experiment. If a farm could prove that it was breeding second-generation Asian arowana—that is, fish whose parents had themselves been born in captivity—CITES would certify the farm to legally export the offspring. Each sanctioned fry would be injected with a traceable microchip and come with a certificate of authenticity, like a work of fine art.

Singapore shipped its first experimental batch of microchipped arowana to Japan in December 1994: three hundred fish, each about six inches long, worth a total of a million dollars. Within two years, by early 1997, CITES had certified nine arowana farms in Singapore, Malaysia, and Indonesia to legally export the species. Kenny's company, Qian Hu, joined the list in 2000, the same year it went public on the Singaporean stock exchange—the first fish farm ever to do so.

In his office, Kenny presented me with a hefty spiral-bound tome entitled "The Most Talk about Fish Farm in Singapore," which contained press clippings tracing his rise as the public face of the newly glamorous fish world. Singaporeans admired him as the filial son who took the family business multinational, a farmer-cum-millionaire, and the charismatic champion of a traditional line of work, which the government once dubbed a "sunset industry." Singapore exports more pet fish than any other country, and Kenny's reign has helped promote a sexy makeover of a historically humble trade—with the arowana ousting the guppy as number one fish.

Now Kenny revealed he was about to take another big step. Shortly before my visit, he had announced that his company would invest $4 million in a research center dedicated to engineering "pedigree dragon fish." Heading this initiative was a scientist named Alex Chang, soon to be awarded a doctorate in fish molecular genetics from the National University of Singapore.

A cheerful, moon-faced man in his late thirties, Alex joined us in Kenny's office to talk about the goals of his research. "We are looking for genes that are responsible for certain colorations, certain characteristics," he explained, "with the ultimate aim of trying to go for a tailor-made arowana."

As he gushed about larger fins and brighter colors, noting how swimming arowanas really look like dragons gliding through the air, his voice betrayed a note of adulation that I hadn't heard in Kenny's. When I asked Alex if *he* believed the fish was lucky, he said he agreed with his boss about the benefits of keeping fish in general but added that the arowana was special: "It has its own charismatic feature—you will see."

Alex had agreed to show me around, since Kenny was leaving in a few hours for an important meeting in Guangzhou. Recently, China had supplanted Japan as the largest market for dragon fish, the species becoming a symbol of upward mobility to the burgeoning middle class. Before turning me over, Kenny told me one last story—about his first pet fish, a goldfish with a broken tail, which he bought for fifty cents when he was seven or eight years old. When he brought it home, his older brothers laughed. "I didn't give a damn, you know?" he said, his expression growing serious. "I still like my fish. I still take care of my fish." A good fish, he decided, was whatever fish you liked best, regardless of marketing or fashion. "You have to form your own opinion about something to enjoy it. You have to define what is beautiful for yourself."

OUTSIDE, THE NOONDAY sun had grown hotter, and more visitors were arriving via free shuttle buses congregating at the outdoor café. Most of the warehouses were off-limits to the public, but the one behind the café was devoted to retail. Rows upon rows of water-filled tanks contained a few kinds of fish I recognized, such as goldfish and guppies, and many more I did not. Alex pointed out small round fish bred for spinal deformities that caused them to puff up like balloons, and others that had literally been tattooed to look as if they were wearing lipstick.

Just past these tanks a pink neon sign and a sculpture of a dragon fish marked the entrance to Qian Hu's star attraction, the "House of Dragon" arowana gallery. Tinted glass doors opened into a cool cavernous space, where dark blue display cases held tanks illuminated from above. It was a bit like the Crown Jewels exhibit at the Tower of London, except with floor drains, damp spots, peeling paint, and jewels that glared at you while swimming in perpetual circles. Reflecting the minimalist aesthetic with which the species is usually displayed, each tank contained a single arowana and nothing else—no rocks, no plants, no backdrop.

As I approached the closest window, labeled PREMIUM HIGH GOLD CROSS BACK, I tried to take Kenny's advice and form my own opinion of the creature. The fish was about the size and shape of a chopping knife, its back the clean, straight line typical of a surface predator, its body covered in large, metallic scales. "The whole fish looks like a gold bar," Alex said with admiration.

More accurately, I thought, it looked gold-plated, with the leafing flaking off in places, so that a dark, swampy brown appeared to be bleeding through its back and head. Alex explained that the gold speckles sprinkled across this desolate, muddy expanse gave the creature the name *cross back*. Despite being a desirable specimen, the fish didn't strike me as beautiful. It swam slowly around the small, cramped tank, balanced by two pectoral fins that stuck out like training wheels. All its motion was concentrated in its undulating posterior, its fan of a tail. As for its face—well, not its best feature. The "bullet head," as Alex called it, seemed strangely small for the body. The mouth was set in a deep harrumph of a frown like a bulldog's. Later, a friend inspecting my photos would exclaim, "That's one *ugly mug!*"

"See how they actually look attentively at you?" Alex enthused. It was true—the arowana's large eyes seemed to follow me without moving, like those of a portrait in a gallery. The fish did have a strange dignity—I'd give it that.

Despite all the fancy names, there are really only three main varieties of Asian arowana: red, gold, and green. White tanks with white lights held the golden fish. Black tanks with pink lights held the reds. Only one green was on display, beneath a sign that indicated it once belonged to Hong Kong actor Chow Yun-fat.

"Two years ago, if you were to come here, people wanted

red arowana. Now people started to go for golden," Alex said. "It changes like crazy." He observed that recently everyone was after "big finnage."

Upon closer inspection, I realized that the arowana weren't actually alone. A small suckerfish slid along the side of each tank, cleaning the glass. "Do they ever eat the suckerfish?" I asked.

"I think they're not so tasty," Alex said, noting that these particular arowana were fed live prawns every day, though technically they would eat anything that moved. "Except for humans—or anything bigger than themselves."

The strange, almost sacrosanct serenity of the space was broken when a children's birthday party burst through the doors. The kids raced up and down the aisles shrieking with delight, though they didn't appear to be looking at the actual fish so much as the price tags. "Wow!" cried a little girl, "5,188 dollars for that one!" (All the prices ended in the lucky number eight.)

"The really big ones are *priceless*," a boy told her, parroting this line over and over when she didn't acknowledge him. "The really big ones are *priceless*. The really big ones are *priceless*."

AFTER WE STEPPED back into the sunshine, Alex explained that he had gravitated to the business at an early age, having "really wanted to go into something fishy." While studying biology at the National University of Singapore, he dreamed of unraveling the biological mysteries of his favorite species— the Asian arowana. But everyone in academia told him not to waste his time. The fish had a breeding cycle of some three to six years, and to get a doctorate he would need to track at least two cycles. Meanwhile, he kept seeing Kenny in the news. Finally, Alex worked up the courage to approach him.

Kenny liked Alex's idea about studying the arowana but

pointed out a hurdle to their working together. Before Kenny would agree to collaborate, he explained, Alex would have to convince Kenny's third-eldest brother, Yap Kim Choon, who was "not educated and doesn't speak English and swears at you when he thinks that whatever you tell him is bullshit." Alex already knew the reputation of this brother—everyone in the world of arowanas did. "He hate studied men," Alex told me. "If you tell him all the theories, he'll ask you to piss off. He'll say the amount of salt you've eaten is more than the amount of rice you've eaten."

As we spoke, Alex led me outside the public arena, across the premises, into the arowana-exporting area, where a man with salt-and-pepper hair peered into an enormous tank crowded with fish, scrutinizing them so intensely that his nose smudged the glass. Arowana are territorial, and you usually can't keep two in the same aquarium (except a mated pair) or they'll fight to the death. Oddly, however, you *can* put a whole bunch together, at which point they seem to call a truce.

Observing that this man looked like an older, surlier version of Kenny, I guessed that he was third brother Yap Kim Choon, whom everyone called Ah-Choon (like a sneeze). Ah-Choon ran the arowana operation, having reared the fish since the 1980s when Kenny was still a college student in Ohio. To say Ah-Choon was hands-on with his arowanas didn't begin to capture it—he kept the youngest fry in incubators in his house. In China, he was supposedly even more of a superstar than Kenny.

Now Ah-Choon turned and gave Alex and me a quick nod before focusing again on the fish. "He's looking for a certain gold tinge to the scale—trade secrets," Alex whispered.

Instead of approaching Ah-Choon directly about undertaking a genetic study of the Asian arowana, Alex spent two whole years talking shop with him. He explained the science behind the fish medicines that Ah-Choon knew well. They discussed the maddening limitations of breeding—that it wasn't possible to tell the sexes apart much less pair specific fish. Finally, Ah-Choon asked Alex what could be done to improve the situation, and he laid out his proposal. "Kenny came to me and said, 'I don't know how you did it, but you managed to convince my third brother, which is like a miracle.' Now I'm working very closely with him," Alex told me proudly, as we arrived at the door of his brand-new laboratory-to-be.

Following his lead, I slipped off my shoes before stepping into the cool, blue room. "Wow," I said and then realized there wasn't really anything to compliment but the concrete floor. The laboratory was still under construction and almost entirely empty except for signs designating areas for microbiology, pathology, microscopy, and cryopreservation. Before long, however, Alex would have all the stations up and running. The holy grail, he told me, was to figure out how to do in vitro fertilization in dragon fish.

I didn't understand why that would be so difficult. By way of explanation, he waved me out the door, back into our shoes, and toward a blue plastic tub of orange-and-white fish nearby. "Okay, this is a koi, a carp," he said, picking up a wriggly male, its little mouth puckering futilely in the air. He flipped it over, gave it a squeeze, and the fish, well, spilled its seed.

"They are always ready," Alex said with admiration. An Asian arowana, he pointed out, would never be so accommodating. "They are fierce, they're carnivorous, they have these very tough teeth. They produce so little sperm you can't squeeze it out like that. We have tried *many, many* times." His face clouded over at the unpleasant memory.

This meant that breeding dragon fish was, in some ways, rather simple: All you could do was plunk them in a pond and wait for them to do their thing. And wait. And wait. As Kenny had told me, "They're not very horny."

Moreover, they are well-known to be picky lovers—they won't mate with just any other arowana that comes along.

Once they make their decision, a good portion of the fish remain paired for life. Yet their offspring are relatively few. Alex explained that each male has only one functional testis, and each female a single working ovary, which releases a few dozen marble-size eggs. The low fecundity of the species was part of what contributed to the vulnerability of wild populations in the first place.

"Come," Alex said. "I show you the ponds before it starts to rain." The sun had disappeared abruptly, and the sky was heavy with clouds as I followed him down a dirt path to a row of rectangular mud ponds shaded with black netting and lined with Indian almond trees.

"It looks like peace and quiet, but there's a lot of breeding," Alex said, staring into the opaque water. Arowana begin courting during the monsoon season, he told me, and courtship can last from several weeks to several months. During that time, each pair circles nose to tail, swimming near the water's surface, especially at dawn. About one or two weeks before spawning, they start to swim side by side, so close their bodies touch. Eventually, the female releases a cluster of orange eggs, which the male fertilizes before scooping them into his mouth, where he incubates the embryos for about a month and a half, fasting the entire time.

Arowanas in love. It sounded complicated. I wondered what else they might be up to down there. "Is anyone studying the fish in the wild?" I asked, explaining that I had been hoping to see the species in its natural habitat.

Alex told me about the researcher who'd discovered Cambodians eating green arowanas, but she was now in Australia studying frogs. Otherwise, he didn't think so. Little was known about the Asian arowana in nature. "The stock is so depleted," he said, pointing out how hard it was to even *find* the wild fish anymore.

"Does it make you sad?" I asked.

"Not really," he said. "Because of their commercial value, people are breeding them and protecting them like treasures." He pointed out that the Asian arowana was not like the Tasmanian tiger—"Gone, gone, gone."

As long as people were keeping it as a pet, the species would not go extinct in any absolute sense, he argued. Also, at least theoretically an "undo" button existed—the possibility that someday the farmed fish could be reintroduced into nature.

Besides, Alex said, the Asian arowana did still exist in the wild, though its numbers were small. Even the coveted Super Red could still be found in Borneo's great Kapuas River—particularly in a vast inland lake system called Danau Sentarum. "One thing I'm very sure of is that at that lake, even until now, there are endemic populations."

"Is it accessible?" I asked, wondering if I might be able to get there.

"Very *un*-accessible," he said. "I don't know whether there is any boat that goes."

As I stared into the cloudy depths of the ponds looking for the arowana lurking there, I tried to imagine their wild progenitors in a dark swamp a world away. Then I considered the fish before me, their offspring destined to become high-end collectibles, circling aquariums ad infinitum.

Alex wasn't sad about that, but I began to think that I might be.

CHAPTER THREE

The Arowana Cartel

SINGAPORE → MALAYSIA → INDONESIA

The dragon fish is one of the most dramatic examples of a modern paradox—the mass-produced endangered species. Since the 1990s, when the farming program began, CITES has tracked the legal export of some 1.5 million Asian arowana, and that number doesn't include the huge domestic markets within countries breeding the fish.

Many Singaporeans seemed to feel that America was simply confused when it chose to ban a creature so common in captivity. "Before you come here, you think maybe there are one or two like the panda!" said Martin Toh, secretary of Fish Club Singapore, imagining my naïveté. I'd met Martin through the cabdriver who picked me up from Kenny's farm, an arowana aficionado who didn't speak English and struggled to convey his love for the fish. Halfway through the ride, the driver handed me his cell phone, and Martin was on the line offering to meet me for dinner to talk arowana. That evening, a dimpled man in a suit showed up at my hotel driving a van I decided probably wasn't the death mobile of a serial killer because it was decked out in fish trinkets. "Want to get frog?" he asked.

The best stir-fried frog turned out to be across town at a hawker stand in the sort of public housing complex where some 80 percent of Singaporeans live. As we drove there, Martin told me he found the subsidized flats cramped and claustrophobic. He resented that the housing authority did not allow fish tanks longer than four feet.

Life in Singapore was comfortable but tightly controlled: public gatherings of more than five people required a permit, which created problems for the fish clubs. A thirty-five-year-old Exxon account manager with a wife and two kids, Martin had taken up the aquarium hobby as a means of escape. "Everyone's nice," he said, "unless you start saying, 'My fish is better than your fish.'"

With few outlets for aggression, he told me, plenty of quarrels took place about fish. Sometimes these disagreements erupted in foul play. Martin's friend Colin Lim had just lost an entire pondful of arowana, supposedly worth some $30,000, and he suspected poisoning. Later, when I met Colin—a bespectacled engineer with a doleful pout—he confided that he'd found an empty drum of chlorine powder near his pond at the clubhouse. I asked who would want his fish dead, and he shrugged, saying there was no telling. "People knew my fish were going to breed," he whispered, as if this explained everything.

The intrigue surrounding the arowana often made getting straight answers difficult. Martin, for example, wouldn't reveal how many dragon fish he owned, saying only, "I lost count." When pressed, he added, "You're asking a guy how much wealth he has. It's not good!" Rather than show me his own arowana, he took me to see others', introducing me to the owner of an employment agency who claimed to have a hundred, several dozen of which were swimming in tanks in the front window of his office. Because the species was so common around town, not to mention in breeding ponds across the region, Martin argued that it was not really endangered.

In truth, however, none of these pet fish *count*—at least not according to wildlife biologists. "A captive fish is the same as a dead fish from a conservation point of view," Chris Shepherd, TRAFFIC International's regional director for Southeast Asia, told me, arguing that no one was going to put captive arowana back where they came from before selective breeding made that impossible—a bit like plunking a dachshund in the tundra. Shepherd called the commercial farming of endangered species his "nightmare" because it opens a laundering loophole: animals poached from the wild are passed off as farm-bred. In the late 1990s, he investigated the arowana trade and found that many of the farms were smuggling some wild stock.

According to Transparency International, Singapore ranks as one of the least corrupt countries in the world. Yet many of the pet fish that pass through its ports originate elsewhere in Southeast Asia. Some 95 percent of Asian arowana shipped worldwide come from farms in neighboring Malaysia and Indonesia, where labor is cheap, land more plentiful, and corruption prevalent.

A week after I arrived in Singapore, I left the island nation and ventured north across the Straits of Johor into Malaysia to visit the country's largest producer of dragon fish, a company called Xian Leng (pronounced *shee-ahn lang*). Only later did I realize what a coup this was, when a professor at the University of Malaya told me he'd once gone undercover posing as a prospective buyer just to glimpse Xian Leng's farms. He was shocked when I told him the company had sent a driver in a black BMW to pick me up in Singapore.

Actually, they hadn't sent the BMW for me but for a Japanese buyer flying in from Osaka with whom I was invited to tag along. This seemed like an even greater stroke of luck. The top Japanese collectors were reputed to be the most discriminating as well as the most private, many rumored to be yakuza, members of Japan's extensive organized crime syndicates. In my research, I'd watched the 1993 Japanese movie *Fried Dragon Fish*, which tells the fictional story of a female detective who's hired to track down a stolen Super Red arowana, but ends up falling in love with a gangster at the center of the fish-smuggling ring. The couple listen to Mahler as they engage in the taboo of feeding each other rare ornamental fish. I'd have to agree with the reviewer who called the film "cod awful," but it planted certain ideas in my head about the kind of Japanese buyer I might be meeting—dark suit, tattoos, missing pinkies.

Instead, the person who strode into the arrivals hall at Singapore's Changi Airport looked like a teenager. He wore designer jeans and had long bangs that grazed his long eyelashes. I wasn't the only one to be surprised. Xian Leng's driver stared in disbelief, then cracked up, laughing. "How old are you?" he asked.

"Thirty," said the buyer, with apparent indignation. Then he pointed his thumb at me. "Who is her?"

The buyer's name was Satoru Yamamoto; and while he was indeed a connoisseur of high-end arowana, it turned out he wasn't a private collector. Instead, he was the aquatic equivalent of an art dealer, working for Kamihata Fish Industries Ltd., the largest ornamental-fish company in Japan. I learned this as we drove out of Singapore, heading two hours north into peninsular Malaysia's southernmost state of Johor, where the ornamental-fish industry first blossomed back in the 1950s with the trapping of wild fish from jungle streams. Today that jungle is mostly gone—displaced by industrial-scale oil-palm plantations producing the popular vegetable oil used in about half of all packaged supermarket items. In recent years, production of the crop has exploded in Malaysia and Indonesia. To my naive eye, however, the endless expanse of cultivated palms looked charmingly tropical—like a neatly organized wilderness.

By late afternoon, we reached the town of Batu Pahat, where

we met Ng Huan Tong, the founder and managing director of Xian Leng. The press had cast Ng as the Malaysian counterpart to Kenny the Fish, pointing out that both men were sons of pig breeders, and the first in their countries to take their fish farms public. After Kenny's debut in 2000, Ng followed suit, listing his own farm on the Kuala Lumpur stock exchange the following year.

The similarities ended there. While Kenny cast himself as a glamorous playboy who insisted on being called Kenny or the Fish, Ng lived with his wife and six children above his fish shop and was definitely a more traditional "Mr. Ng." Unlike Kenny, he actually looked like a farmer, stocky, with a low center of gravity, thinning hair, and a face weathered by the sun. He had left school at age thirteen, in 1971, to help run the restaurant his family opened after giving up pig farming. In his late teens, he started rearing tropical fish in his backyard and founded Xian Leng after noticing that his arowana were breeding.

In 2002, the year after taking his farm public, Ng sold what was then touted as the most expensive aquarium fish of all time—an albino Asian arowana with red eyes—reportedly fetching 20 million yen (about $150,000 US) from an anonymous Japanese buyer. Back then, Ng's operation faced only a handful of serious competitors. Now it vied with more than eighty farms in Singapore, Malaysia, and Indonesia, leading to losses from aggressive price competition. If Kenny was always smiling and extolling his company's slogan—"We bring fun to you!"—Ng's resting expression was one of worry. Even when he grinned, he grimaced.

That evening over fish-head curry at a courtyard restaurant next to his shop, Ng admitted frankly that he regretted going public. "I thought life was not enough challenge," he said, quaffing a beer. Now he was under constant pressure from the board. "Other businesses, after listing, can double, triple, et cetera. Agriculture cannot."

That's why, he explained, he needed the American market to open—the reason he'd decided to let me tour his traditionally closed farms. The United States, which imports the most pet fish of any country in the world, has staunchly refused to support the commercial trade of a species listed as endangered. Ng wanted to show me the plenitude of arowana multiplying in his ponds.

At eight the next morning, he picked up Satoru and me in a blue jeep, and we set off past green hills, pink houses, and a gaggle of schoolgirls in matching brown head scarves who hopped off the side of the road as we passed. Eventually, we turned onto a dirt path that ran through an oil-palm plantation. Hidden deep within the trees sat a fifty-acre farm encircled by steel fencing. Inside, ninety-six ponds carved into the red earth gleamed like amber in the morning sun. Some five thousand arowana—parent stock—lurked in the brown waters, leaping up to catch live cockroaches that workers tossed them by the handful. Ng credited this exclusive diet with his fish's trademark sheen. He bred the roaches himself in a strangely sweet-smelling shed, where they teemed in vast quantities in open wooden bins.

Now he ushered us into a much-larger warehouse that contained rows of tanks, each holding a single arowana not much bigger than a Swedish Fish candy, though far less colorful. The fish were supposedly golden—the star breed of Malaysia—but looked silvery gray with faint yellow fins.

The tricky thing about buying a baby arowana is you can't tell what it's going to look like grown up. The fish don't "blossom" into their adult coloring until they hit sexual maturity, which takes about three years for a gold and six for a Super Red.

Satoru's job was to look at the seemingly identical young fry and augur their futures. His forehead beading with sweat, he began by rejecting any unforgivable flaws. "Ventral fin is broken," he said, pointing to one and drawing an X on the tank with a red marker.

"Bump on head," he said of the next.

"This one has an eye problem. Left eye uneven."

"Bottom lip juts out."

"Split fin."

At last, he stopped to admire a sixth. "This fish has strong golden color," he said, then shook his head ruefully. Faint white spots indicated that its tail had once been broken. In the end, he marked ten tanks with the word *BUY*. Ng looked dismayed. Outside, he told me in a low voice that Satoru's company usually ordered at least a hundred.

Looking out across the tranquil farm, I couldn't help feeling a bit disappointed myself. Given all the rumors about smuggling, robberies, and organized crime in Malaysia, I had expected more of a Wild West full of fish bandits. Instead, Ng spent the afternoon driving us through the peaceful countryside, Satoru nodding off in the front seat. We visited not just arowana farms but facilities breeding goldfish and guppies as well as all kinds of strange and unexpected species. At one point, Ng pointed to a small black pancake of a fish in a shallow pool and told me fondly, "Stingray baby. Just born this morning. Never seen the world."

THE AQUARIUM TRADE moves a greater number of species— more than seven thousand—than any other industry, according to Ornamental Fish International. Marine fish kept in saltwater tanks are nearly all captured wild from the ocean, but freshwater fish make up about 90 percent of the trade. Of those, all but 10 percent are now bred on farms, primarily in Asia.

It seemed to me like an unequivocally good thing that the vast majority of pet fish are no longer collected from the wild. Critics, however, point out that farming removes an important economic incentive to protect nature. Malaysia, for example, has shown little concern with preserving the small jungle streams that made it a center of the aquarium trade in the first place.

When I realized I was now in arowana country—the original habitat of the fish—I asked whether the species still inhabited local rivers. "No more arowana," Ng told me. "Very few."

On one of our stops, he introduced me to Raymond Cheah, the president of the Federation of Malaysia Fish Association, a gentle, bright-eyed guppy enthusiast, who invited me to join him on a day trip to collect wild fish and aquatic plants in the surrounding countryside. A few days later, I met up with Raymond and a visiting Finnish aquarist named Tor Kreutzman, a towering man with white hair and a dark beard who always spent his holidays in pursuit of exotic fish.

"I was here in the eighties, nothing but leeches," Tor told me, as we drove along a well-paved highway. "About ten, twelve years back, this road was still a single lane." He remembered there had been a lot of mouse deer the size of small dogs.

"Now, not many," said Raymond, behind the wheel of the truck.

"They've been *eaten*," Tor said.

At breakfast, Raymond had told us we could obtain whatever delicacy we wanted around here: tiger meat, elephant, sun bear. "It's illegal," he said, "but if you know the owner . . ." Later, he mentioned that at least once a year a local was eaten by a tiger. A few years back, someone had been swallowed by a python.

The jungle, however, was not easy to find. We passed elephant-crossing signs but no elephants. Monkeys jostled the trees by the side of the road, crowding a thin strip of forest as illusory as a movie set. Eventually, we ended up traipsing on foot through an oil-palm plantation, where Tor used a small net to catch Siamese fighting fish from a muddy ditch. Raymond said there used to be green arowana in a nearby river—but not anymore. "Because of the money behind it," he said. "Very, very crazy price."

The other way that farming a fish can backfire for its wild brethren is by increasing its popularity as a pet. Rather than tamping down the craze for the Asian arowana, commercial breeding of the species seemed to have stoked the flames, igniting a spate of thefts beginning in the early 2000s. For a while, I wondered if this crime wave belonged to the past. Neither Ng nor Kenny the Fish claimed to know anything about the robberies. "Stealing a fish is not as simple as stealing a piece of jewelry," Kenny had told me.

Raymond, however, tipped me off that no one wanted to talk about the dark side of the trade. "Most of the arowana breeders are very secretive," he said. "They tell you half, and then they stop." To show me the real story, he took me to meet his friend Tony Tan, who owned a farm called Wo Long Asian Dragon Corporates. Nestled inside yet another oil-palm plantation, the place had all the charm of a military compound: nothing but a high concrete wall topped with two rows of coiled barbed wire. A watchtower hovered over the entrance, and a riot of ferocious barking greeted our arrival, though the dogs, caged till nightfall, were somewhere out of view.

As the gate creaked open, Tony dashed toward our truck, a lean, tense figure, dark from the sun, with a cigarette hanging out of his mouth. A former car mechanic, he had taken up fish farming "Chinaman style," as he put it, meaning from scratch and on his own. At first, he raised goldfish and koi. But those fish made little money, so he switched to arowana—though it hadn't been easy.

Pointing to three workers mending a hole in a green wire fence, he told me that two weeks earlier, a thief had scaled his farm wall and cut through the fence, netting thirty arowana in the night. That was nothing, he said, compared to three years back when ten armed men appeared at the farm in the evening.

They muscled their way past the gate, touched a pistol to his temple, then tied him up, along with seven of his workers. The thieves spent four hours catching arowana and packing them into a truck. At one point, they took a break to cook and eat eggs from Tony's chickens. When they finally left at two in the morning, they made off with 60 large fish and 402 fry. That was the second armed theft at the farm. In a decade of raising arowana, Tony said he'd been robbed eight times.

I asked who he thought was taking the fish. "They are professionals," he told me, speculating that the goods were probably smuggled north over the Thai border. He had reported five of the thefts to the police, but little came of it. "I am very frank. I say how it is. Our government cannot do anything. They know only this." He rubbed his fingers together to indicate money.

According to Tony, the most frightening incident occurred some five months before my visit, when fifteen men arrived at the farm midafternoon, threatening to kidnap and torture him if he didn't sign bank papers approving a transfer of half a million US dollars. He signed. After the men left, he stopped the transaction. The young leader of a local gang was arrested and spent four days in jail before being released on bail to await trial.

That evening, when we joined Tony for a dinner of deep-fried snakehead fish, I asked if he ever ran into the assailant around town. He pointed past the open door of the restaurant to the dark street beyond and said, "His truck is outside."

"That's why I don't like to keep arowana," Raymond whispered. "At least I can sleep at night."

THE DARK STORIES that swirled around the dragon fish were dizzying and difficult to reconcile with the larger picture of the hobby. The collectors I met certainly seemed harmless. They called themselves arofanatics, sported temporary fish tattoos, and took to chat boards to express their passion for the species through the liberal use of emoticons. Many spoke of fish-keeping as a great stress relief, bringing peace and calm.

One night before I'd left Singapore, Martin from Fish Club had taken me to his father's flat, which was decorated with glossy red paneling in an ode to the Super Red. We sat at the kitchen table with the lights off, sipping orange soda and staring at an illuminated tank. It was important to Martin that I experience the fish properly, that I "get" what he saw in the arowana.

"Assume it's one a.m.," he instructed me in a soft voice. "You're stressed. You've been scolded by your boss. . . . Now just watch the fish." His hand swam hypnotically through the air. "Watch the movement of the tail."

I did find myself strangely transfixed by the rippling motion of the sinuous body. But I couldn't imagine myself watching a fish for hours each night, as many hobbyists claim they do. Then again, I do not fit the profile of a typical aquarist much less an arofanatic. The aquarium hobby is traditionally male dominated, and keeping dragon fish especially so. "Arowana is for men," Ng Huan Tong told me on his farm. "Women don't like such a creature." Another Malaysian described the species as an outlaw: "It's a very aggressive fish. It's a cowboy—like Billy the Kid." He shot his fingers in the air and then called the arowana the "Ronald Reagan of fishes."

Reagan seemed to come up a lot, actually—maybe because of his own cowboy image, or because the Asian arowana first shot to fame during the Reagan years in the 1980s. (Incidentally, Reagan was the only American president to introduce a First Fish into the White House—a goldfish named Ron who lived in a tank with a presidential seal.)

The other eighties icon mentioned frequently was Rambo. I noticed the standard pose that farmers—and even some serious hobbyists—strike when showing off an arowana in a pond resembles nothing so much as wielding a bazooka. Commonly featured on business cards and promotional materials, the stance involves wading waist deep into the water, catching the fish (not an easy task, best accomplished wearing heavy gloves), clutching its rump under one armpit while using the opposite hand to clamp shut its mouth, then brandishing it cross-body like a weapon.

All this seemed so goofy and benign—mere playacting at being a tough guy—that it was hard to take seriously the stories about the underbelly of the arowana trade. At the same time, however, a genuine sense of menace was attached to the fish. A Malaysian member of parliament once reported a "death threat" in the form of a text message featuring a photograph of a dead arowana. "I fear for my safety now," he told the press.

Rumors of fish-related violence abounded. The worst case I could confirm was the brutal stabbing death of thirty-one-year-old Chan Kok Kuan, the former welder turned aquarium-shop owner in northern Malaysia. In the five years since the murder, no killer or killers had been arrested. When I took a three-hour bus trip north from the capital city of Kuala Lumpur to Chan's hometown of Taiping, I found his shop shuttered. Later, once I'd connected with the family, the father, Chan Ah Chai, explained that after discovering his son's mutilated body, he'd insisted the place be closed forever, saying "The fish business is such a dangerous and risky thing."

Taiping was merely twenty miles from the lake of Bukit Merah—the fabled habitat of the golden arowana. Stretching over seven thousand acres, the vast expanse of blue water was supposed to be one of the last places in Malaysia where the wild fish still lived. What I found when I got there, however, was not a pristine wilderness but a giant resort featuring water slides, carnival games, and fast-food restaurants on well-lit piers. In a nearby village, one fisherman told me he'd caught many arowana here in the nineties, but now they were gone: "No more fish." Recently, the host of a Malaysian angling show had secretly planted a farmed arowana in the lake to pretend to catch it for the cameras.

I was beginning to feel that the real mythical beast was not the well-branded dragon fish swimming in tanks around the world but the wild arowana itself. It was like Big Foot or the Loch Ness Monster—a creature people talked about and believed in but had not actually seen. Of course some *claimed* to have seen it. Martin from Fish Club told me he'd spotted an arowana in Bukit Merah while on holiday there. When pressed, however, he laughed uncomfortably and amended the story: he'd seen a glimmer of gold in the water.

It seemed to me that if the wild arowana did exist, it probably survived in a far more remote and less developed place. As I stood eating McDonald's amid swaying cattails, watching the morning mist rise from the lake, I found myself pondering not the golden arowana that once thrived in these waters, but rather the legendary Super Red, the "Ferrari of fishes," from Indonesia. I felt the Super Red was, in a sense, the *real* dragon fish, the one that had propelled the market for the arowana into the stratosphere, appealing to the Chinese mania for red. ("That's why Pepsi cannot penetrate into Asia," Martin had explained, reasoning that a blue can never stood a chance against a red Coke.) Alex Chang, Kenny the Fish's head research scientist, had told me it was still possible to encounter the Super Red leaping from the water in Borneo, a name that bore the unmistakable whiff of adventure, conjuring primordial rain forests filled with fantastical creatures. I swallowed the last of my fries and decided the fabled island would be my next destination—to find the wild fish.

I SHOULD'VE REALIZED Borneo is not a small place. It is almost a mini-continent—the third-largest island on earth and a prize gem in the Malay Archipelago, the chain of more than twenty-five thousand islands strung between Southeast Asia and Australia. For most of the last millennium, Borneo was isolated from the rest of the world. The first Europeans to set foot there in 1521 were part of Magellan's pioneering round-the-world voyage (though Magellan himself, killed in the Philippines, didn't make it that far). The boggy interior, however, proved virtually impassable. Not until the nineteenth century did Dutch and British imperialists begin to carve up the island in earnest.

Today Borneo is divided by three nations: Indonesia, Malaysia, and the oil-rich kingdom of Brunei, which is smaller than a single cattle ranch it owns in Australia. The Super Red comes from Kalimantan—Indonesian Borneo—which comprises the majority of the island, roughly the southern two-thirds. In my travel guide, Kalimantan appeared as a vacant gray shadow, its only tourist attraction being an orangutan reserve to the far south. Not mentioned at all was Pontianak, a gritty port city of half a million people nestled in the vast marshy delta of the great Kapuas River, where I expected to find the Super Red.

Armed pirates were said to roam the river, raiding the arowana farms clustered on its banks. "There are so many farms there—so many farms," Alex Chang had told me. "Some farms are almost half the size of Singapore."

The farmers, I reasoned, would know where to find the wild fish because they had plucked the species from nature in the first place, creating the international trade. Yet I had trouble making any local contacts, just as an ichthyologist at the National University of Singapore had warned me. "These farms are heavily guarded. It's almost like controlled by mafia," he'd said. "If you know the correct people you are safe, but if you don't, you are risking . . . life, maybe."

Eventually, a trader I'd met in Singapore put me in touch with a young man named Willy Sutopo, whose father had founded one of the oldest arowana operations in town, and who agreed to help me get the lay of the land. From Malaysia, I flew some nine hundred miles south to Indonesia's capital city of Jakarta, where Willy spent half his time. He met me at the airport, a hip twenty-eight-year-old Chinese Indonesian wearing plaid shorts, flip-flops, and a red Nike shirt. He had spent a few years studying accountancy at Purdue University in West Lafayette, Indiana; and as we sat waiting for our flight to Borneo, he told me Pontianak wasn't so different from West Lafayette. Both towns had one cinema and one mall. Both had gossips. "Everyone knows everyone," he said. "It's not so nice."

Willy slept the entire ninety minutes as we flew northeast across the Java Sea, the water beneath us dotted with tiny lakes of land. When the west coast of Borneo finally rose into view, I looked down to see the muddy snake of the Kapuas River winding through the city of Pontianak, a jumble of low buildings with red and blue roofs.

In Indonesian mythology, a *pontianak* is the vampiric ghost of a woman said to have died during pregnancy who returns to exact revenge by clawing out men's sex organs. In the 1700s, the first sultan of Pontianak gave the town its name after he thought he saw such a phantom haunting his palace grounds. Locals take the ghost seriously and watch out for pale women with long hair—a description I unfortunately fit to a T. In the bathroom of the small, chaotic airport, an unsmiling lady in a head scarf snapped a point-blank photo of me washing my hands and then raced out the door.

Around the same time Pontianak got its name in the 1700s, gold miners began arriving from southern China. As in other parts of Southeast Asia, they formed *kongsi*, or trade associations, which were part secret society, part self-governing state. Today the town's population is majority Chinese, many the descendants of these original settlers, and the arowana business falls within the tradition of the *kongsi*. The farmers are the wealthy big shots around town.

A black SUV with tinted windows was waiting for Willy and me at the airport. As we sped along streets crowded with motorbikes, I caught glimpses of the main strip: men smoking on low plastic stools outside coffee shops; a factory producing mosque-toppers, where piles of crescent moons glittered in the sun; and a giant billboard featuring a Malay man with a mega-watt smile advertising a get-rich-quick scheme that involved the Super Red.

"If you walk into any of the houses, it's more likely than not you'll find they have an arowana at home," Willy told me. "It's like the native pride." A lot of locals made money by rearing the fish to a certain size and then reselling them. Our first stop was a small white house belonging to Willy's uncle, who kept half a dozen arowana in his wok-lined kitchen.

One of the fish was such a dark crimson that it was almost black. "My uncle's favorite pet," Willy said, as we sat on the linoleum floor admiring the strange creature. "He's the darkest arowana that ever lives." The color, which had emerged unex-pectedly, was puzzling. Usually such a fish was blind, camou-flaging itself according to its perceived environment of total darkness. But this one could see.

That arowana alter their coloration according to their envi-ronment makes it possible to fake certain color varieties. Put a green fish in a white tank, and after a while it starts to appear gold. Place a red one in a black tank and it turns even redder. I wondered what the fish looked like in nature without all this manipulation. I had been on the road for two weeks now, visited five arowana farms, and seen many, many fish in tanks and ponds—but not one that "counted" in an ecological sense.

Though the last thing I felt like doing was visiting another farm, Willy's place lay half an hour up the Kapuas River, and I was eager to escape town and get closer to the species' natural habitat. As we drove through green fields and rice paddies, however, Willy told me no wild arowana could be found this far downriver anymore. If they did still exist—which he thought was possible—it must be in the distant headwaters, the vast network of seasonal lakes and intermittently flooded forests known as Danau Sentarum. Willy himself had never been to Sentarum and had no desire to go. He had taken a Japanese buyer half the way there, and the experience was not to his liking: bad roads, intolerable sleeping arrangements, inedible food. He wasn't a fan of roughing it.

For years, his family's farm had been highly manicured. Recently, however, Willy's father had decided that it would be more beautiful left to its own devices. On our arrival, I noticed weeds growing wild on the edges of the dugout ponds. We spent the morning watching the old Super Reds—some of them nearly our age, in their late twenties. These were the original fish, the first generation that his father had brought back from Sentarum while working as a logger penetrating the interior. Now, however, they were acclimated to a domestic life. A worker tossed live frogs in the water, which the arowana leapt up to catch with their bony tongues, their glistening flanks more orange than red now.

For security, the farm was encircled by four nested walls, and the ponds were set back about three hundred yards from the river. As the call to prayer sounded in the distance, we walked along a grassy field to the water's edge, looking out through barbed wire at passing barges. Willy told me that his next-door neighbor, who had an exclusive contract supplying Kenny the Fish, had recently been robbed. "I heard from quite a reliable source, but I cannot verify," he said. "It's quite secretive."

I asked if we could venture onto the water—to at least see where the arowana *used* to live—but Willy said it wasn't safe here. Later he took me to hire a paddleboat that an oarsman steered through town, passing between banks crowded with a chaotic jumble of wooden houses on stilts. Old men bathed in the water, and little boys swam in their underwear, hooting as we passed. A trio of ladies shading themselves with pink umbrellas crossed our path in a boat.

As we glided upriver under the fierce equatorial sun, I squinted in the direction of the remote headwaters. The Kapuas River is the longest in Indonesia, winding some seven hundred miles, roughly the distance from New York to Chicago. Across all those miles, I felt the dark, far-off swamp that spawned the Super Red tug at me like a magnet. I wanted to get to the wild place—to see a fish that counted as part of the natural world.

I spent the rest of the week trying to figure out how to reach Sentarum and growing increasingly frustrated. It simply didn't seem possible. No regular boats went that far. A flight to a town in the interior left once a week but was often canceled, and you could easily get stranded due to rain. Even more problematic was the language barrier: few people in town spoke any English, and the situation was bound to be much worse outside Pontianak.

On top of having no one to travel with, no way to communicate with locals, no swamp know-how, and zero fishing experience, I couldn't afford to get marooned indefinitely in the heart of Borneo. I had to fly back to Singapore, some four hundred miles away, where the entire fish world was converging for a massive extravaganza, the biggest global event in pet fish, which promised to be—as advertised—"The place you would not wish *not* to be at!"

CHAPTER FOUR

Aquarama

SINGAPORE

A few days later, I stood blinking in the razzle-dazzle under the bright lights of the Suntec Singapore Convention & Exhibition Centre.

"Why you don't have dragon fish in America?" asked a small, clipboard-carrying man, wearing silver glasses and a gray shirt with epaulets, who'd been darting around the cavernous exhibition hall before stopping short before me. His name tag identified him as Dr. Ling Kai Huat of Singapore's Agri-Food & Veterinary Authority, chief judge of the Aquarama International Fish Competition. The aquatic equivalent of the Westminster Dog Show, the event had been held in Singapore every other year since 1989, kicking off a four-day trade show. This year forty-five judges, hailing from about half that many countries, were gathered to assess more than a thousand show fish competing for top prizes.

Gurgling tanks held "fancy" goldfish with trumpeter cheeks and telescope eyes—some so plump they swam like astronauts leaping on the moon. Siamese fighting fish with billowy tails occupied small cubic vessels separated by cardboard blinders, which could be slid out to induce territorial displays. Beady-eyed catfish sat still as stones or vacuumed the glass, exposing grotesque sucker mouths. Cool blue marine tanks contained corals, clown fish, and strange skinny pipefish poking their heads out of rocks. Then there were guppies, and guppies, and more guppies, all with flowing tails like flamenco skirts in every conceivable hue and cut.

At the center of the exhibition, circling their aquariums like poorly synchronized swimmers, were the arowana, grouped by color, one contender to a tank. As always, the two main categories were the classic Super Red and the Malaysian golden, which had recently become especially trendy. "Japanese like gold," Ling told me. "Chinese like red." Selecting a winner, he said, came down to personal taste, "just like fashion for ladies' wear."

Though the judging was technically top secret, closed to the public and press, Ling offered to show me how to assess a dragon fish: Begin by inspecting its color, body shape, body size, and finnage. Long fins are of the moment, as is a hunched spine causing the fish's head to dip down like a spoon. Look out for disqualifying conditions. No matter how nice a fish is in other regards, if it's PLJ, meaning it has a "protruding lower jaw" (essentially an underbite), it cannot win. Likewise, if it suffers from "droopy eye," a mysterious affliction whereby arowana in aquariums look down too much and can't look back up, the fish is no good.

"Like this one definitely out," Ling said, stopping to point out an arowana staring at the ground. "Looks so sad." He made an exaggerated frown. (Such an afflicted fish, however, need not be a lost cause. "Dr. Arowana," Singapore's premier mechanic-turned-fish-repairman, pioneered the use of diamond-cutting tools to fix up arowana eyeballs in the early nineties. Two decades later, there's now a cottage industry of fin jobs, jaw tucks, and piscine eye lifts.)

Ling went on: Swimming style is important, as is personality, the more aggressive the better. Then there's uniqueness. Like a great work of art, a great fish must embody a degree of conformity, adhering to an established set of technical rules, while simultaneously defying on some level the very aesthetic tradition of which it partakes. In other words, you want a fish that is at once recognizable and novel. Something familiar but different. Something *new*.

One vehicle of novelty is mutation. A truly notable arowana might, for example, have conjoined fins or, on occasion, two heads. The most sought-after mutation by far is albinism, which occurs very rarely in nature. A true albino materializing on the market used to cause the stir of a long-lost Vermeer. This year, however, in a move set to shock the industry, someone had entered a whopping ten albinos in the exhibition. It was unclear who owned the fish, as the tanks were marked merely with numbers; but whoever it was had thought to hire a private guard, a young man standing watch with a pistol on his hip. I stopped to look at the row of pure white arowana swimming in ghostly parallel, like marble replicas of fish that had been dead for millennia, and asked the guard how much they were worth. He shrugged and said, "To someone who doesn't like them, they're worth nothing."

Ling's tour ended in front of a Super Red about a foot and a half long, gliding unhurriedly in short, dignified laps. It had a classic knife shape, its mouth set in a deep fleshy frown, recalling an old man without a chin. While Ling couldn't reveal which arowana had won—the results were still being tallied—he lavished praise on this fish, contrasting its apparent health and red sheen with its less impressive neighbors.

The arowana did look exceptionally red, with cherry cheeks to match its scales; though in truth, it was hard to perceive its real color, because the tank was lit by the hot-pink aquarium light in which red arowana are routinely displayed. The eyes of the fish reflected this glow. They gleamed large with black pupils, unmoving, unblinking, so that I was reminded of a story by the Argentine writer Julio Cortázar about a man who becomes obsessed with a primitive aquatic salamander called an axolotl on view at the Jardin des Plantes in Paris. He stares at one of the creatures until he finds himself trapped within its consciousness, watching his own body walk away from the aquarium. "Above all else, their eyes obsessed me," he explains. "I felt a muted pain; perhaps they were seeing me, attracting my strength to penetrate into the impenetrable thing of their lives. They were not human beings, but I had found in no animal such a profound relation with myself. The axolotls were like witnesses of something, and at times like horrible judges."

I stood watching the Super Red until Ling indicated we had to move on. By the end of the day, its tank would be plastered with a crimson ribbon bearing the title GRAND CHAMPION. For the fish, this meant four days of paparazzi harassment. For its owner, it could mean tacking an extra zero onto the value of his pet. All this was nothing, however, compared to the bragging rights. At an arowana competition in Jakarta, a man had recently shown up wielding two guns and demanding to know why his fish hadn't won. One British judge told me he would never let himself be assigned the dragon fish—because he didn't want his head cut off.

THE FOLLOWING MORNING I slipped into the opening ceremony of the trade show just as the first speaker—none other than Kenny the Fish—was taking the stage, camera flashes glinting off his trademark smile. "May I use a metaphor?" he asked the crowd, eliciting a few anticipatory snickers. Fish metaphors were Kenny's specialty. I knew this because my attempt to download his book, *The Rise of an Asian Entrepreneur* ("Just $103.95! $19.95!"), had yielded a preview chapter entitled

"12 Fishy Pointers to Your Business Success!!," a guide to what entrepreneurs can learn from different fishes. (The lesson of the arowana, for example, was to evolve a robust internal organ structure.) Now Kenny leaned into the mike. "Time is like a fish you catch with your bare hands," he said. "It slips away!"

A medley of groans and giggles rose from the crowd. This year Aquarama had drawn more than forty-three hundred industry insiders from eighty countries, ranging from the United States and Japan, to Congo, Kazakhstan, and Papua New Guinea. Kenny spoke to the fortitude of the industry in the face of the recent global economic crisis, pointing out that aquarists who lose their jobs don't typically go home and kill their fish. It was a strange speech. For those with long memories, it was probably cold comfort. On the last day and a half of the show, when the doors opened to the public, some fifteen thousand visitors would descend on the exhibition—a far cry from the fifty thousand that the inaugural Aquarama drew some twenty years earlier.

Hard hit by electronics and video games, serious fish-keeping seemed to be going the way of now-quaint hobbies such as model building and coin collecting. Old-timers spoke in elegiac tones about glory days past, when dedicated aquarists kept fish in backyard sheds, sought obscure species, worked hard to breed them, and were proud members of thriving aquarium societies.

Paradoxically, however, the industry has continued to grow, going mass-market and focusing on "bread-and-butter" fish that American retailers such as Walmart now sell at prices cheaper than ever before. Globally, aquarium fish exports were worth just $21 million in 1976. Today they've increased to the $300 million range, according to the United Nations Food and Agriculture Organization. Add in shipping, retail markups, and accessories, and the industry probably clocks in at around $15 billion. The big bucks are not in the fish themselves so much as their *stuff*: tanks, food, lighting, filters, pebbles, plastic castles. This is true of the modern pet trade in general: the money is in the accessories. If pets themselves have become commodities, we are now buying commodities for our commodities.

Such splurging, however, is not without precedent. The last people to invest heavily in fish accessories were those masters of conspicuous consumption the ancient Romans. In the early first century BC, fish ponds called *piscinae* came into vogue among the ruling elite, who carved cavernous grottoes into seaside cliffs, adorning them with fountains and waterfalls, sculptures and walkways, even elaborate marble dining pavilions. The Romans' all-time favorite aquatic pet, that unapologetic phallus of a fish the eel, began to command outrageous prices—to say nothing of the *jewelry the fish wore*. Emperor Claudius's mother is said to have attached earrings to hers (never mind its lack of external ears). Eventually, pet fish became so closely associated with decadence that the great statesman Cicero used the term *piscinarii*, or "fish-pond owners," to refer with disdain to the rich. "They are such fools," he wrote. "They seem to expect that, though the Republic is lost, their fish ponds will be safe." In the end, Cicero was right. Rome fell and the *piscinae* crumbled, their pampered eels slithering out to sea.

THOUGH PEOPLE HAVE been keeping ornamental fish in pools and bowls for millennia, the modern aquarium had a specific genesis in mid-nineteenth-century England, where natural history was a full-on obsession. The story begins, oddly enough, in 1829, when the London doctor and amateur entomologist Nathaniel Bagshaw Ward came across a moth chrysalis, which he brought home and shut in a jar. Months later, he found the moth hadn't hatched, but a fern had sprouted from a damp bed of mold. Ward was puzzled that the fern had grown *without once being watered.* He had made an important discovery: that plants may thrive in a container sufficiently sealed to prevent the escape of water. The moisture that the greenery transpires condenses on the glass and drips down like dew to the soil, forming an all but closed system. This invention, the "Wardian case" or terrarium, spawned a fern craze of epic proportions. By the late 1840s, no fashionable Victorian drawing room was complete without its "vegetable jewelry," as Shirley Hibberd, the male Martha Stewart of his day, observed. These "plumy emerald green pets glistening with health and beadings of warm dew" constituted the flatscreen TVs of the mid-nineteenth century. People sat in their parlors and stared at leaves growing behind glass.

Soon the Victorians grew bored of watching ferns and wanted something more. Help was on the way. Spurred by a relatively new body of scientific knowledge, the chemist Robert Warington got the idea to turn the Wardian case upside down, fill it with water, introduce aquatic plants, and add *fish*—thus inventing the Warrington case (which somehow gained an extra *r*), soon to be called the aquarium. While fish had been kept in glass bowls before this time, the water required constant changing or else the fish died. What Warington theorized in a paper he presented to the Chemical Society in 1850 was that if plants and fish occupied the same glass-enclosed environment, the plants would give off enough oxygen to support the fish. This deceptively simple concept was far more revolutionary than it might at first seem, since the basic chemistry of respiration and photosynthesis hadn't been understood until the late eighteenth century.

Like the mania for those lotto-predicting flowerhorns that raged through Singapore in the early 2000s, the British aquarium craze of the 1850s was a passionate but short-lived affair. It coincided with the relatively new practice of taking seaside holidays, and droves of Britons flocked to the shore, buckets and trowels in hand, to cull the inhabitants of tidal rock pools. Soon sea anemones were the hot fashion of the moment, with the writer George Henry Lewes joking that they were only slightly less troublesome to keep than a hippopotamus. He had a point. Sustaining captive sea life was no easy task. By the 1860s, "nine out of every ten aquaria were abandoned," according to the naturalist John George Wood, "and to all appearance the aquarium fever had run its course, never again to appear, like hundreds of similar epidemics."

But it wasn't over elsewhere. Just as the tomato originated in the Andes but took on new life in the hands of the Italians, the aquarium may have been invented in England but came to thrive in Germany. While the British craze centered on ocean life, the Germans were the first to embrace freshwater fish. In 1856, the schoolteacher Emil Adolf Rossmässler published the article "Der See im Glase" ("The Lake in a Glass"), promoting the hobby as a means of democratizing scientific knowledge. The aquarium was, after all, based on a rudimentary understanding of what today we call an ecosystem. It was a world in miniature, an idyllic picture of perfect balance in which the respiration of fish (oxygen in, carbon dioxide out) and the photosynthesis of plants (carbon dioxide in, oxygen out) created a self-supporting microcosm.

It was also a fantasy. Measurements done in the twentieth century would reveal that the plants in an aquarium release a trivial amount of oxygen, at least compared to the amount rapidly absorbed into the water from the air. Likewise, the amount of carbon dioxide that plants remove from aquariums is also inconsequential. In other words, the whole aquarium craze of the 1850s was based on a misunderstanding. What enabled fish to survive in captivity was not the perfect balance of plant and animal life but rather the shape of the modern aquarium itself. Thanks to Britain's abolition of the glass tax in 1845, large panes of glass could be used to build rectangular vessels that increased the size of the water-air interface and enabled the rapid absorption of oxygen into the water, and more important, the slow transfer of CO_2 into the air. No perfect "balance of nature," no steady state, exists—not in an aquarium, and not in nature itself. There is only flux.

If at first the aquarium hobby provided a way for ordinary people to explore the wild close to home, the desire for more exotic fish eventually spurred expeditions into terra incognita. In the late nineteenth century, German businesses established collecting stations in ever-more-remote areas of Asia, Africa, and South America. As German sailors brought back new species, and German ichthyologists provided most of their scientific descriptions, knowledge of the natural world kept growing apace.

Out of this culture of discovery, at the turn of the twentieth century, emerged a gentleman naturalist from Frankfurt named Adolf Kiel, pioneer of the aquarium hobby, "father of water plants"—and grandfather of a legend I was about to meet.

AS KENNY THE FISH addressed the crowd at Aquarama, an American in a Hawaiian shirt poked me in the shoulder. "That's Heiko Bleher," he whispered, and I turned to see a bearded man toward the back of the room who stood out like a construction cone. He was wearing a wide-brimmed Australian bush hat and a bright yellow parka over a black T-shirt printed with a perfectly round tropical fish. Slung over his shoulder was a leather satchel, as though he'd just stepped off a plane. "Ask him about being arrested for biopiracy in Brazil," the American said mischievously.

I had already heard much about Heiko, a former fish trader turned famous ichthyological explorer. He liked to say he was from "all over the world," but he was born in a bunker amid the ruins of 1940s Frankfurt, not far from the site where his grandfather ran one of the largest ornamental-fish and aquatic-plant farms of the day. At sixty-four, Heiko technically lived outside Milan but still spent most of his time traveling to remote locales in search of new fish species, of which he claimed to have discovered—or rediscovered—some five thousand. (In the parlance of the trade, "rediscovering" a fish means finding a species known only as a shriveled museum specimen, recording its occurrence and behavior in nature, and ideally introducing it to the hobby.)

Heiko was often called "the Indiana Jones of the tropical fish industry," which he seemed to take as an insult, believing he could best any fictional character. This year, however, a high-profile scandal had tarnished his superstar image. The previous fall, in 2008, he had been arrested in Brazil, along with his brand-new twenty-nine-year-old wife, thirty-four years his junior, for allegedly attempting to carry fish preserved in formalin and alcohol out of the country without a permit. He held that he was targeted for his outspoken opposition to destruction of the rain forest. But his reputation for egoism was such that his having spent three and a half days in what he pronounced "the worst prison on the planet" seemed to have inspired more schadenfreude than sympathy.

When I introduced myself, Heiko told me he had just flown in on a red-eye. "Before yesterday I was in Canada, and the weekend before I was in Poland, and the weekend before in Turkey. Every seven weekends I was collecting." He handed me his card, which featured a drawing of a fish with a snout like a chain saw. "The large-tooth sawfish," he said, explaining he had discovered a ten-foot-long specimen of the creature in 1982, while diving at night in a crocodile-infested lake in northern Australia. He maintained it was the largest freshwater fish recorded in the twentieth century.

I mainly wanted to talk to Heiko because I'd heard he had ventured deep into Borneo to Sentarum, the home of the Super

Red. I was just asking him about this when he was distracted by the sight of an elderly Japanese man perched on a stool in front of a prominent exhibition booth. It was Shigezo Kamihata, the eighty-three-year-old chairman of Kamihata Fish Industries Ltd., known for its highly regarded koi pellets. Kamihata's face was spotted with age. He wore a seersucker suit and dark aviator glasses, through which he regarded Heiko with an expression of bemused forbearance.

"Kamihata-san is explorer number *two* in the world," Heiko told me cheerfully, by way of introduction.

The old man made a small bow and took my hand. His skin felt thin as rice paper. Like Heiko, he was a third-generation fish trader, his grandfather having pioneered koi breeding in 1877. Later in life, Kamihata began collecting wild fish, making extravagant expeditions on chartered planes, together with a team of divers. Now he reached across the booth and gave me a book entitled *The True-Life Jungle Adventures of Kamihata in Search of Tropical Fish*, the cover of which featured him standing on a rock, arms akimbo, in the midst of a roiling river.

Heiko promptly took the book out of my hands to show me, with great pride, the first line of the first chapter: "Mr. Heiko Bleher, the German explorer, called me one day and said, 'I'd very much like to show you the unexplored splendors of the Amazon.'" Only later would I notice the heading atop this page: "Heiko Pushes Our Nerves to the Limit." The chapter describes a 1989 expedition during which Heiko fails to pick up Kamihata at the airport, books an unsatisfactory hotel, womanizes, sleeps in late, and proves to be so spectacularly hairy that searching for ticks on him is like "finding a needle in a haystack."

What interested me most, however, was the penultimate chapter of Kamihata's book: "Searching for Wild Asian Arowana in Borneo." In 1993, he too had made an expedition to Sentarum, the remote wellspring of the arowana, in pursuit of the Super Red. Explaining that the swamp was inhabited by the Iban, a tribe once notorious for headhunting, he showed me photographs of an elaborately tattooed man and a woman holding up a giant crocodile skull. Even in the early nineties Kamihata had been unable to find the Super Red in its natural habitat. Fishermen reported that whole months would pass with no sign of the species. He feared that by now the arowana was probably in danger of extinction.

Heiko shrugged. "There's so much water. It's impossible to say—especially because it's so far away. People don't go there to collect except very rarely." He said he would take me to Sentarum himself if he didn't have back-to-back expeditions planned in Australia, the Philippines, and Iran. "You come with me to the Amazon in Colombia," he offered instead, describing his plans to visit what he characterized as one of the last true wildernesses on the planet.

Then he reached into the leather satchel slung over his shoulder and brought out his own newest book, which he showed Kamihata. The cover featured a black-and-white portrait of a blond woman with an elaborate updo, framed by sketches of Amazonian flora and fauna. "This is the history of my mother in the fifties," he said, flipping to a spread of photographs— a montage of tribal women with pendulous breasts, a man waving a shrunken head on a stick, and a vat of skulls and bones. He placed his finger on this last image. "These are the missionaries eaten by the people we lived with."

Kamihata leaned in to have a closer look, raising his eyebrows. "Where?" he asked coolly.

"Mato Grosso in Brazil. I lived there six months as a boy."

If the venerable Kamihata was a latecomer to jungle exploration, Heiko had practically been born to the life. After World War II, his mother, Amanda Flora Hilda Bleher, daughter of Adolf Kiel, "the father of water plants," opened a mail-order pet business in Frankfurt, importing exotic animals from around the world. In those years, she acquired the "It fish" of the day, a perfectly round discus from the Amazon—then the most expensive aquarium fish in the world—for an exhibit in the monkey house at the ruined Frankfurt Zoo. She was mesmerized by the perfect geometry of the creature, its flat, leaflike body and sweet pursed lips. So when a clairvoyant Gypsy later advised her to go to Brazil, into her mind swam the rare Amazonian discus she had seen only once. In 1953, having divorced her husband, Amanda packed up her four children and set off to pursue the fish in an uncharted section of the Amazon rain forest, commonly known as "the green hell." Heiko, the youngest, was nine when they were stranded in a leper colony and held captive by a Nazi on the lam.

That first expedition lasted nearly two years, after which the Bleher family went home to Frankfurt. In 1958, however, they returned to Brazil, this time for good, and Amanda established a fish-breeding facility and aquatic-plant nursery outside Rio de Janeiro. In 1962, at the age of eighteen, Heiko moved to the United States to study ichthyology at the University of South Florida in Tampa, where he worked at several of the area's largest fish farms, making a reputation for himself as a fish wrangler able to coax wild-caught species into breeding in captivity. Two years later, he returned to Rio, where he opened his own aquarium-fish exporting company and began crisscrossing the Amazon on collecting trips. At the end of 1964, when he was twenty years old, he discovered the first new species to be named after him—*Hemigrammus bleheri*, the rummy-nose tetra, which he maintains is the second-bestselling tropical fish of all time. It was one of ten species, including an entire genus, that would eventually bear his name.

As I followed Heiko into Aquarama's fish exhibit, I noticed that he gravitated toward the Amazonian discus, the creature his mother had sought more than a half century earlier. After decades as the reigning high-end fish, the discus had been knocked off its throne by the arowana—though not in Heiko's heart. In his 2006 book, *Bleher's Discus: Volume I*, he insisted, "The discus cult knows no bounds, and this fish will undoubtedly always be number one. . . . There have been times when fish have changed hands for $10,000 or more. Some people have even sold their house to buy a particular discus—losing wife and children in the process."

He was not, however, impressed with the selection at Aquarama. "The fish they have here are very poor. The first-prize discus, I would throw him in the garbage," he said, after stopping to examine the grand champion, a golden albino with red eyes. He complained that the judging was atrocious: "One of the discus judges from England knows discuses as much as I know about life on planet Pluto!"

It bothered Heiko that the show discus no longer resembled the subtle creature his mother once fell in love with—a perfectly round, pale green fish with black vertical stripes. Over the years, mostly at the hands of Asian farmers, the discus had been bred into what Heiko deemed a garish assortment of colors, with names like leopard skin, snakeskin, checkerboard, blue diamond, and pigeon blood. It was increasingly difficult to find the original strains anymore.

Heiko's disdain for the fish reflected a deeper rift that has rent the aquarium world in recent decades. Selectively bred, highly domesticated varieties have grown increasingly popular, especially in Asia. Meanwhile, a subset of aquarists, particularly Europeans, have become exclusively interested in wild-type fishes, turning up their noses at faux "red fish," as they call them. If one side sees art in fish, the other sees nature—and wants it tampered with as little as possible.

This philosophical schism is evident in the kinds of aquariums championed by the opposing groups. Heiko, for one, advocates the "biotope aquarium," which aims to re-create a single ecological niche, bringing together the flora and fauna of one river or lake—never, say, combining African fish with Amazonian plants. The biotope stands in contrast to the high-profile "nature aquarium," as conceived by the Japanese aquarist Takashi Amano, dubbed the Sage of Aquariums by the *Wall Street Journal*. Nature aquariums are lushly planted tanks, highly coiffed, with Zen-style arrangements of rocks, and small fish darting around as a "supporting cast." They aim to perfect nature. To Heiko, nature *is* perfection. It cannot be outdone.

For this reason, when Heiko noticed all the attention being lavished on the ten albino arowana—strange, artificial "monsters" that they were—he glowered at the fish, and the large crowd they had attracted. I, however, was curious to learn the story behind the fantastical creatures and went off in search of an explanation.

I FOUND THE man responsible for the mysterious cluster of snow-white fish, a thirty-four-year-old Chinese Malaysian fish farmer named Alan Teo, standing in front of the display, looking exhilarated and exhausted. He wore gold glasses, gold cuff links, and a gold pen tucked into a tailored shirt with a mandarin collar. Even his dark hair, slicked into a neat part, glinted under the bright lights of the showroom. By his own estimate, Teo had slept a total of seven hours over the past five days. "I feel very nervous," he confessed, frowning at the fish. "The water is very bad over here, and the light is wrong."

A police detail had escorted the albinos to the exhibition, and the privately hired guard stood watch to prevent anyone from adding poison to the tanks. "As the one and only breeder of the world's most expensive fish," Teo said, "you make a lot of enemies." He couldn't shake the memory of what had happened six years ago to the first-place discus—it disappeared overnight.

According to Teo, the albinos represented the culmination of a clandestine, years-long experiment. Once upon a time, he had started a modest arowana farm in Malaysia and found that his fish, by chance, produced a single albino offspring. Word of the birth reached an Indonesian coal tycoon—a billionaire who kept a private zoo strictly for albino animals deep in the jungles of east Borneo. This powerful figure traveled to Teo's farm and sat for hours, in silence, watching the little arowana swim back and forth. The next day, he announced that he wanted to bankroll a breeding program. Teo would have to wait three years for his albino to grow old enough to reproduce, giving birth to offspring that appeared normal but carried the recessive genes for albinism. Only after another three years could he breed the original arowana with its children in the hope of yielding more albinos. In the meantime, under the strict orders of his benefactor, Teo was to stop selling fish.

"Our priority goes to celebrities and royalty," Teo said, explaining that the albinos were now for sale by invitation only. Fewer than ten had been purchased so far, all by anonymous buyers, each in a different country. A prominent member of the Chinese Communist Party had recently bought one for $300,000. Another had sold to a Las Vegas casino baron, who requested that it be shipped to Canada, where, unlike in the United States, the species is legal. The only such fish in Taiwan belonged to a plastics magnate who made his fortune manufacturing toothbrush bristles.

"Some people think it's just rumor, but it's true," Teo said of his tale. "This fish has let me see lots of things I'd never dreamed of." He held up his hands, which had long, delicate fingers, and demonstrated how they'd trembled the day he installed an albino in the private chambers of the sultan of Johor—a man notorious for having allegedly murdered a golf caddie who snickered when he missed a hole.

I asked Teo if I might meet one of the owners of the albinos, and he considered this. The sultan of Johor was on his death-bed. His son was a possibility, but in addition to being even more of a wild card—infamous for having shot a man dead in a nightclub—he wasn't that into fish. Then Teo got an idea. He said one albino owner might, in fact, be willing to grant me an audience, and excused himself to make a call.

In his absence, Heiko Bleher rejoined me, sizing up the albinos, his bloodshot eyes nearly as red as theirs. "Three hundred thousand dollars," he scoffed. "I wouldn't pay thirty cents for that fish." Then he said I would never get to the bottom of anything in the arowana world. "Specific to this fish, no one will tell you the truth."

When Teo reappeared, however, he had come through with an introduction to one of his clients. A few days after Aquarama ended, I flew some two thousand miles across the South China Sea to make one final stop—to meet for myself the owner of a six-figure fish.

CHAPTER FIVE

The Dragon's Den

Late on a Thursday afternoon, I stood waiting on a lawn as pristine as AstroTurf outside a large white house with a grand portico in the suburbs of Chiayi City, which lies in the southwestern plains of Taiwan. May had given way to June, and the soft rays of the sun burnished the surrounding lotus fields. I stopped to peek inside an aviary built around a tree, which contained a pair of velvety blue birds with black heads and long tails—Taiwan blue magpies, the national bird of Taiwan, illegal to keep as a pet in the country, being kept as a pet nonetheless.

Before Aquarama, I'd set my sights on meeting an elite Japanese collector, but hadn't been able to turn up any leads. For example, when I asked the Malaysian arowana farmer Ng Huan Tong who had bought his record-breaking albino for 20 million yen back in 2002, he claimed he *forgot*. "Big boss," he said, shaking his head as if the name were on the tip of his tongue. "Arowana lover."

I wasn't the only one who wondered if such shadowy characters actually existed. "You never know about these arowana stories," a fish judge at Aquarama warned. "They pay huge sums. But nobody knows who bought it. Nobody knows where it is."

Now, however, I finally had an introduction to meet such a top-tier collector—a tall man in his sixties who eventually came striding across the lawn, his welcoming smile revealing a strong set of profoundly white teeth. It was Su Wen Hung, the so-called toothbrush-bristle tycoon, president of the successful manufacturing company Rai Hsing Plastic Machinery Works, and owner of the only albino arowana in Taiwan.

"He also makes the hair of the Barbie," whispered my companion, Stephanie Lee, a young journalist who worked for the Taiwanese fish magazine *AquaZoo*, which was publishing a story on Su's arowana. She and her boss, Osmond Chao, a ponytailed snake enthusiast, had picked me up the previous evening at the Taipei airport, and we'd driven three hours south to Chiayi City. Over the past twenty-four hours, we had stayed at a hotel that Su owned and gone to see a new house he was constructing, which would feature a floor-to-ceiling tank for the albino. Mostly, however, we killed time, waiting to meet the man himself.

It was actually quite fitting that I had ended up in Taiwan, the very place the cult of the arowana first arose in the 1980s before spreading throughout Asia. What sparked the craze is debatable. In Chinese culture, pet-keeping was long associated with the upper class—the most pampered pets in history probably being the Pekingese court dogs, which suckled at the breasts of human wet nurses in the Forbidden City. Come the Cultural Revolution of the 1960s and '70s, however, China banned pets as "bourgeois." In the 1980s, when China's government was still describing pet-keeping as "the product of a sick capitalist society," the Taiwanese discovered the Asian arowana and suddenly created the ultimate status-symbol fish. Was Taiwan thumbing its nose at the Chinese Communist Party? Or celebrating its Chinese cultural heritage? Or was it just that Asians really love fish? As a Hungarian fish biologist working in Singapore later told me, "You look at their relationship to anything that comes from the water, and it's very different from ours—it's so much more emotional."

Emotional, however, the toothbrush-bristle tycoon was not. He had less to say about his arowana than I'd heard aquarists tell me about their $2 guppies. In Southeast Asia, the aquarium hobby is often called "playing fish"; and Su had many other playthings to occupy his time, such as the Harley-Davidson motorcycles and model helicopters crowding his spacious foyer. Portraits of prize-winning pigeons lined the walls, and trophies from German shepherd competitions gleamed in display cases. He bred the pigeons on the roof. The dogs he kept caged in his own private kennel, where a terrible barking rattled the bars. "He wants his animals to be perfect," Stephanie translated. "He likes things that are very rare and precious, that no one else can own."

Su told us he had paid $150,000 US for the albino arowana—half the price quoted at Aquarama for the fish sold to the Chinese official, though still among the highest I'd heard. He said this was no big deal; he had a pigeon worth twice as much and a dog worth twice that. As his wife passed through the foyer laden with shopping bags, he paused to wave hello, then whispered that he'd left off a zero when he told her the cost of his fish.

To my surprise, the albino wasn't even on the premises. While awaiting completion of Su's new home, it was living in the back den of an aquarium shop across town, an elegant gallery space where arowana swam in windows set in wood-paneled walls. I'd seen the prized fish the night before, spending what I deemed a polite amount of time admiring the other-worldly, waxen creature, its only hint of color a faint, orangey glow on the tips of its fins like the first rays of dawn. Had it really cost Su $150,000?

Verifying who paid what for which fish is like authenticating the inflated prices that art dealers routinely report—all but impossible. The dedicated fish press isn't exactly hard-hitting. Even I found myself torn between the twin impulses of watchdog and cheerleader, skeptical of the high prices I was so keen to confirm. During a moment of miscommunication, when I thought Stephanie had told me Su's albino cost $50,000 rather than $150,000, I grew livid to have traveled so far to see such a cheap fish. Like other forms of thrill seeking, from BASE jumping to cocaine, chasing luxury fish suffers from the tolerance effect, with ever greater extremes needed to achieve a proper high.

More interesting to me than the albino itself was Guo Xianming, the owner of the aquarium shop housing the fish. He had recently toured the private albino zoo in Borneo belonging to the coal magnate bankrolling Alan Teo's breeding program. Over tea served on a glossy tree trunk of a table in the front of the shop, Guo showed me photographs of snow-white monkeys and translucent alligators next to a pitch-black coal mine that appeared to have swallowed the animals' pigment.

I was still dismayed not to have ventured into Borneo's interior myself. Before coming to Taiwan, I had once again tried to figure out how to get to Sentarum, but it simply didn't seem feasible. A nervous travel agent warned me about Islamic terrorists hiding out in the jungle—members of Jemaah Islamiyah who might seize the occasion of Indonesia's upcoming presidential election to kidnap a Westerner for attention. "Besides, there are headhunters," he said.

Though I didn't believe that indigenous tribes still practiced head-hunting, I didn't know what to make of the terrorist bit. When I e-mailed the security officer at the US embassy in Jakarta, he didn't seem worried. "The area's extremely remote," he acknowledged, but he hadn't heard of anyone, American or otherwise, being kidnapped in west Borneo.

Oddly enough, I *did* know of such a case, though it had

nothing to do with Islamic terrorists or headhunters—it had to do with arowana. In 1998, Yoichi Kawashima, the thirty-one-year-old son of a Japanese fish importer, was reportedly abducted in Pontianak and held ransom for ten days by a group of men wielding machetes, handguns, and grenades. I'd read about the incident in a Japanese newspaper, which didn't print the name of the accused ringleader, though I eventually learned he was a prominent Indonesian arowana farmer. In Taiwan, it seemed as if I saw his stern scowl everywhere I turned. His photo was on the wall of Guo's fish shop. There it was again in Su's foyer. The man supplied their Super Reds.

All this gave me the disconcerting feeling of being out of my depth. And it should have made me happy to be going home—leaving behind the murky realm of might-be fish gangsters. As Martin from Fish Club had told me at Aquarama, "You're trying to see for yourself what's real, but you cannot. Because part of the truth is covered up. So whatever you hear, it's bits and pieces."

He was right. I was sick of struggling to make sense of rumors. For that matter, I was even sicker of looking at arowanas. I decided that I actually found the fish kind of ugly, with its gnarled visage, petulant pout, and wormlike barbels. Their tanks almost always seemed too small, and I felt bad watching them negotiate each lumbering U-turn. Despite myself, I worried they were horribly bored, "condemned infinitely to the silence of the abyss, to a hopeless meditation," like the axolotls in Cortázar's story.

As I sat in the wood-paneled den of the Taiwanese fish shop, watching the wan, consumptive-looking albino circle its tank, I had all these familiar thoughts. But there was something else too. It occurred to me that the fantasy being enacted with the arowana was not just about seeking some essence of wildness, something aggressive, and even a bit scary, as I'd come to believe. It was also about the opposite—control. Installing an arowana in your living room may start out as a way of introducing a dark, fearsome beast into the safe, civilized cage you inhabit; but it quickly morphs into a means of conquest. It's about taking a primordial predator, camouflaged to lurk in black water, and draining its ferocity, until you're left with a creature as translucent as the bristles of a toothbrush, as bleached as Barbie's hair.

I considered how far the fish in front of me had strayed from its natural origins and wondered if its progenitor still survived in the swamp. The question gnawed at me for reasons I couldn't fully explain. I just knew I had to go back to Borneo. I had to find the wild arowana, to see for myself what was true. The following night, I canceled my flight home and began my own lumbering U-turn.

IN THE FISH world, all roads lead back to Kenny the Fish. So it should come as no surprise that the person I finally found to venture with me to Sentarum was Kenny's exclusive supplier of Super Reds. Hery Cheng, a twenty-nine-year-old Chinese Indonesian, ran one of the largest arowana farms in Pontianak, which his father had started in 1980. The day after I left Taiwan and flew back to Indonesian Borneo, Hery led me high up a watchtower to look down on his massive expanse of concrete ponds next to the Kapuas River. Some five walls separated his farm from that of his neighbor, Willy Sutopo—the breeder who'd accompanied me on my first trip to Borneo. I remembered that Willy had mentioned a rumor that the farm next-door—Hery's—had recently been robbed. When I asked Hery about it, he claimed it wasn't true, laughing and raising his eyebrows. "So people are talking to you," he said.

That Hery could laugh off such a question set him apart. Hosting a journalist had made Willy nervous. By contrast, Hery stood out as a bastion of calm in a sea of paranoia. He had sleepy features, a perpetually bad case of bed-head, and always seemed to be taking a slow drag on a cigarette. He had taught himself English with the same shrugging nonchalance that he displayed when he agreed to travel with me to Sentarum. He made it clear that he thought there was zero chance we'd actually find the wild Super Red. He was, however, game to try.

Our first stop was to meet a leathery-skinned farmer named William Albertos Tomey, the man who had accompanied Shigezo Kamihata—"explorer number *two* in the world"—to Sentarum in 1993. This was the expedition I'd read about in the book Kamihata had given me at Aquarama, which contained a photograph of a vast lake adorned with a tiny island. A king had supposedly been buried there, and the locals avoided the place at night, believing it to be haunted by his ghost. Kamihata, however, had insisted on using the island as a base camp anyway; and one of his guides claimed to have been awoken by a vision of blue fire rising from a rock on which the face of an old man appeared.

Now, all these years later, Tomey drew us a map of the swamp from memory, laughing about how he'd gotten Kamihata lost there. He circled a dot in the center of the page—the strange little island—and said it was here that we'd find the native Iban who could take us to the habitat of the Super Red.

According to Kamihata, the head of the tribe had invited his party to stay three nights, promising them their pick of a different woman each evening. "I imagined they wanted our blood—our DNA to be exact," Kamihata wrote of the isolated community, "like a stud horse service for humans." He said he declined the offer, telling the patriarch, "Our only aim is to visit the spawning grounds of the wild Super Red."

After giving me the hand-drawn map, Tomey broke the news to Hery that a local arowana competition scheduled for July had just been canceled over worries that the upcoming Indonesian presidential election might destabilize Borneo. Certainly, the Chinese farmers had been targeted during periods of political instability in the past. The worst episode had corresponded with the collapse of the three-decade Suharto regime in 1998, when ethnic Chinese came under attack across Indonesia, where they represent a tiny fraction of the population but control a very large portion of the country's wealth. Like many, Hery and his family fled to Malaysia during the worst weeks.

I had heard about the crisis—about how military tanks rolled in to guard the biggest arowana farms, which seemed to add to the dark mystique of the fish. I didn't, however, realize what else was going on in Borneo at the time. Hery explained this to me after we left Tomey's farm and drove to pick up supplies at a supermarket.

While nearly half the people in Borneo belong to various Muslim groups, an additional third or so are Dayak, a loose term for dozens of indigenous tribes such as the Iban. In the late 1990s, fighting broke out between the Dayak and the Madurese, Muslims whom the government had, over thirty years, relocated to Borneo from the poor, arid island of Madura. Before the conflict finally reached a standstill in 2002, more than a thousand people were killed, and another 180,000 displaced. "I think the Dayak even eat of the Madurese flesh," Hery said. He must have seen the horror on my face because he gave me a reassuring grin and added: "No problem!"

I was still trying to wrap my mind around what he had told me when we reached the supermarket, which was gigantic and fluorescent-lit with wide, gleaming aisles. A little boy raced down one of them and leaped into Hery's arms—the youngest of his three children, out shopping with a nanny. Hery hadn't seen him in a while, having recently separated from his wife. He'd told me how hard that had been, and how expensive it was to pay for private school, as all the Chinese did, since the public schools in Indonesia taught Islam. These seemed like fairly relatable problems, but as I stood in that oddly familiar, sanitized setting, picking out shampoo with Hery, what I found myself worrying about were cannibals.

I wasn't sure I believed Hery's tale of modern-day human flesh-eating. Later, however, I'd confirm he was right. As part of the ethnic violence, Dayak gangs had indeed beheaded their victims and consumed their livers, reviving an ancient practice among headhunters. A Dutch priest who witnessed the chaos told the British journalist Richard Lloyd Parry:

> You have to try to understand the position of the Dayak people now. They are ignored by the government. They have no political role. No one in the key positions, no people of influence in the army. . . . All they have is land, land that has been theirs for thousands of years. Now the government appropriates the land. . . . The timber companies come, other commercial concerns. . . . It is one thing, one thing inspired it all: powerlessness.

EARLY THE NEXT morning, Hery picked me up in a gray 4x4, and we set off up an old gold-mining road running along the Kapuas River. For the most part, the road wasn't bad—certainly not bad enough to prevent the driver Mr. Hussein from barreling down steep inclines and around blind bends, narrowly skirting motorbikes, and slamming the brakes just ahead of ditches, small children, and trucks loaded high with the black, prickly fruits of oil palms.

Not so long ago, most of Borneo was virtually impassable. In the eighties and nineties, however, loggers chopped down its rain forests at an astounding pace, exporting more timber than Latin America and Africa combined—primarily to wealthy countries, such as the United States and Japan. Then came the oil palms. Between 1990 and 2005, the amount of land cleared for oil-palm plantations across Indonesia tripled. All this deforestation and development rendered remote places easier to access, leading, in turn, to a spike in the illegal wildlife trade.

Today, in theory, the nearly three-hundred-mile drive northeast from Pontianak to the upper Kapuas River should only take about twelve hours. But if the front tire goes flat, and there are "tolls" (men with knives and a collection basket who've rolled a log in the road), and then the truck gets stuck in the night behind a stranded vehicle—it's a two-day trip. We spent the night in the town of Sintang at a hotel Hery had been hoping to avoid, where he said he'd once counted some thirty cockroaches in his room. I slept mummified in a mosquito net I couldn't figure out how to hang.

As of yet, the primordial rain forest I'd expected to find remained out of sight. The next morning, as we continued on, I asked Hery about a building with an ornate, sloping roof that looked to me like a temple rising from a clearing. "It's for getting the driving license," he said. A DMV.

I stared out the window and tried to imagine the bygone days when the first Westerner made it to the top of the Kapuas River—George Müller, a Dutch official who was promptly beheaded by the Dayak in 1825. To be fair, things might have gone differently had he not arrived directly from coercing a Malay sultan into signing a treaty recognizing Dutch sovereignty. The sultan had put a price on Müller's head.

The village of Semitau, the final outpost on the edge of Sentarum, was the last indigenous territory to fall under imperial control, as late as 1916. For most of the next century, it was a poor place with wooden shacks and bamboo piping. Then came the arowana boom and a flood of money, bringing asphalt and plumbing. When Hery and I arrived there late in the afternoon, we were surprised to find that a small inn called the Arwana Borneo Hotel had just opened to put up fish traders.

Two large photographs of Mecca hung in the foyer, and a store across the way sold coffee thick as crude oil.

We decided to spend the night before venturing upriver, and that evening, Hery took me to see the man who'd sold his father his first arowana back in the 1970s. Haji lived in a house built on stilts over the swamp. He was in his late fifties, had a pencil mustache, and wore a blue sarong. "There are still some arowana in the wild, but not around here," he told me through Hery. "From the wild, the fish is bigger, more beautiful, more red." It had been four or five years since he had seen one himself.

When I unfolded our hand-drawn map, Haji poked his finger into the dot at the center. "Pulau Melayu," he said, naming the island, "Iban." We would find the tribe on the small speck of land—it was just as Kamihata had said in his book.

THE NEXT MORNING, Hery and I set off upriver in a small motorboat belonging to Haji's son-in-law. Soon we were passing through a tangle of secondary growth, a new forest in its infancy. Striations in the clay along the shoreline indicated that the water level was low. Dry season was coming. Now and again, we passed the floating houses of seasonal fishermen, who would soon be migrating downriver to work on oil-palm plantations.

Hery was dismayed that he hadn't been able to find a single Super Red for sale from local fishermen. Acquiring fresh stock from the interior had recently taken on heightened urgency, as the arowana in Pontianak were producing much smaller broods. The problem was almost certainly mercury poisoning, a byproduct of illegal gold panning, which contaminated the water that the fish farms used. There wasn't enough gold remaining in the riverbed to yield much money—certainly not as much as arowana—but no one had been able to stop the practice. Hery said that two days before our trip an angry mob had burned down a police station to protest the authorities' interference.

In the early afternoon, our boat rounded a bend and Sentarum suddenly spilled out before us, a vast expanse the color of tea. To the north rose the pale blue mountains of the Kapuas Hulu Range. Just visible at their base, there was no mistaking it—Pulau Melayu, the island in Kamihata's book, as tiny as a period punctuating the horizon.

For a while, it seemed as if we might not get there after all. The water level was lower than expected, and our boat kept getting stuck in the mud. But we waded the final stretch and at last climbed up the steep, rocky shore. If the island looked like an idea set down in the world, it was even more surreal for being entirely deserted. The Iban who would supposedly take us to the Super Red were nowhere to be seen.

"I'll put in a nose ring, and you can take a picture of me," Hery offered, trying to cheer me up.

We clambered up a short trail to the top of a steep central hill, where we took in the view of the surrounding lake reflecting the blue sky. I had the eerie sensation of standing on the pupil of an eyeball.

Come to think of it, the island itself felt strangely creaturelike. Locals said it had magical properties, that you could poke a dead stick in the ground and it would start to grow. In a clearing in the brush, I spotted a large rock scrawled with graffiti, which I recognized from Kamihata's book—the grave marker of the old king who supposedly haunted the island.

When Hery and I heard laughter coming from the trees, however, it did not have a supernatural source. Rather, we turned to find that a sunbaked Malay fisherman wearing a pair of blue briefs had come striding up the trail and stood cracking up at the unexpected sight of us. As Hery explained what we were doing here—how I had come from New York in search of the arowana and wanted the Iban to take me to it—the man laughed even harder, his eyes growing wet with tears.

Once he regained his composure, he introduced himself as Ali and announced with good cheer that we had managed to pick the absolute worst time to visit Sentarum, the two-week period when the Iban were essentially stranded in the village where they lived a few hours upstream. The water was too low to take a boat there, but too high to walk.

I squinted in the direction he was pointing, toward an out-lying shore, and said it didn't look *that* far. Ali insisted it was farther than it looked. He had tried to walk the distance himself and thought he would die. He mimed this out, leaning down and wheezing, hooting as he lifted each foot out of invisible mud.

When I asked how many hours the trek would take, his face fell abruptly. He turned to Hery and said I couldn't be serious—there were *crocodiles*. He pointed to the crocodile breeding ground, saying the big ones were crossing back and forth.

Looking out across the murky water, I wondered what else lurked beneath the surface that I could not see. According to Ali, he had caught fifty arowana two or three years ago, but now none were left. Then he changed his story and said he caught fifty every January.

"To keep asking will make no sense," Hery said. "I think he is not telling the truth."

As we headed back down the trail to leave, Ali gave me a friendly whack on the back, and I turned to see his extended hand bearing the gift of a sun-dried fish, stiff and whiskered. When I told him I couldn't accept the whole thing, he tore it to pieces, handing me the belly.

At the bottom of the hill, we found Ali's family, two women and three children, had puttered up in a small fishing boat. The older woman, clad in a DKNY T-shirt and flip-flops, carried a basket from which she proudly extracted an eagle, feetfirst, its hackles catching on the weave. When I asked why it didn't fly away, she laughed and made a motion as if to slice off her arm at the elbow, indicating that its wings had been clipped.

For some time, we stood there chewing on the salty fish, watching this pet eagle limp and cry. It had chestnut plumage with a head as white as the string that trailed from its foot. Long after I got back to New York, I would be as haunted by the memory of this once-wild bird as I was by the fish I could not find.

PART II

THE EDGE OF THE KNOWN WORLD

CHAPTER SIX

The Living Fish

NEW YORK

Fish idolatry is nothing new. In 1905, a farmer searching for bat guano to use as fertilizer climbed deep into La Pileta Cave in southern Spain and discovered a prehistoric drawing that for its size and realism left him breathless. The chubby, five-foot-long "Great Fish," thought to be a halibut, was sketched on the cavern wall some eighteen thousand years ago. By the dawn of civilization, the Sumerian pantheon included the fish-like sage Adapa, believed to have climbed out of the Persian Gulf on human legs. When the Assyrians rose to power around 2,500 BC, they worshipped Dagon, a "sea-monster, upward man / And downward fish" (in the later words of John Milton). But it was the ancient Egyptians who left behind the most material evidence of their reverence for the aquatic set.

In the British Museum's gallery of ancient Egypt, you might overlook the unassuming relics of these long-ago devotions amid the crowds flocking to see the mummy of Cleopatra and the Rosetta Stone. But in a quiet corner off to the side, I once stood before a small glass display case transfixed by what appeared to be a heavily bandaged thumb nestled inside a six-inch box. A sign beneath it read:

MUMMY OF A FISH IN A PAINTED WOODEN COFFIN
PTOLEMAIC OR ROMAN PERIOD, AFTER 305 BC

The ancient Egyptians embalmed mass quantities of fish, often in their own sarcophagi, which they buried in dedicated piscatory cemeteries. When I made an appointment to view the substantial collection in the museum's cavernous basement archive, a young archaeologist led me past storage units crammed with human mummies to an aisle where he pulled out a drawer full of snakes and fish. They didn't look like much: just bundles of flaking linen in glass jars. To the ancients, however, they were divine liaisons, purchased by pilgrims to convey prayers to the gods.

Like so many other shifts in worldview, the idea that fish rank beneath us in a "great chain of being" can be attributed to the ancient Greeks, specifically to Aristotle. Among the great philosopher's contributions to Western thought—his logic, ethics, metaphysics, and political philosophy—his writings on fish aren't usually mentioned. Yet Aristotle actually thought a great deal about them. As the first biologist, the originator of the scientific study of life, he may have influenced the modern world most profoundly. Today his longest work is also his most neglected—the ten-volume *Historia Animalium* or *History of Animals*. In it, he attempts to organize living creatures based on complexity, ranking them on a *scala naturae*, or "ladder of life."

Aristotle wrote *Historia Animalium* during his time on Lesbos, the island to which he retreated after the death of his mentor Plato in 347 BC. Having been passed over to assume leadership of the Academy, the younger philosopher could scarcely have picked a more beautiful location to contemplate the natural world free from the politics of Athens. A sparkling turquoise lagoon cut deep into the island, and Aristotle spent his days in the shallows, studying mullets, sharks, gobies, and rockfishes. Inspecting translucent cuttlefish embryos at various stages of development, he marveled over the mystery of inheritance. He reasoned that more than just matter passed from parent to child, but rather some organizing agent—what today we know to be DNA.

Aristotle's teacher Plato had been a creationist, maintaining that God made the cosmos and then turned sinners into animals as punishment for bad behavior. The flighty became birds, the foolish became snakes, and the most ignorant of all became fish. By contrast, Aristotle didn't believe in a divine designer, instead holding that nature itself was sacred and unchanging. In a stirring passage known as the "Invitation to Biology," he writes that studying the heavens is less profound and important than inspecting the lowliest creatures on earth. "For in all natural things there is something marvelous," he observes. "If someone has considered the study of the other animals to lack value, he ought to think the same thing about himself as well." Here was the clarion call of modern biology—one that would echo unheeded for two thousand years as science moldered through an age of superstition.

After classical times, mythical monsters and sea serpents larded the pages of natural histories until the sixteenth century, when Renaissance thinkers began churning out encyclopedic compendiums as part of a broader cultural awakening. Yet while Aristotle had striven to organize animals into related groups, these scholars merely listed them—and their lists were short. Even as late as the early 1700s, European naturalists knew of only a few thousand species of animals, a mere 150 of which were fish. That number, moreover, was obscured by convoluted nomenclature resulting in redundant descriptions of the same species under different names.

In the traditional narrative of history, a single Enlightenment luminary strode forth to impose order upon chaos. The Swedish taxonomist Carl Linnaeus divided each natural kingdom (animal, vegetable, and mineral) into classes, then subdivided the classes into orders, genera, and finally species. Most significantly, he invented the system of binomial nomenclature—the idea of giving each species a name consisting of just two Latin words—still used today. The first word defines the genus (such as *Homo*, meaning "man"), and the second names the species (such as *sapiens*, meaning "wise"). Up to this point, species often had long and unwieldy names. For example, the Atlantic cod was called *Gadus dorso tripterygio, ore cirrato, corpore albicante, maxilla superiore longiore, cauda parum bifurca* (cod with three dorsal fins, a mouth with a barbel, long upper jaw, the tail barely bifurcated). Linnaeus pruned this back to *Gadus morhua*.

Quite comfortable on the subject of his own brilliance, Linnaeus believed he was chosen by God to reveal the divine order of Creation, modeling the hubris that would come to characterize the modern conquest of nature. "No person has ever proved himself a greater botanist or zoologist," he wrote in one of his five autobiographies. "No person has ever so completely reformed a whole science, and created therein a new era." If this sounds a bit heavy-handed, maybe that's because Linnaeus knew that someone else lurked in the shadows who could perhaps make such claims.

Surprisingly, I first learned about this forgotten figure when I visited the Linnean Society in London, the world's oldest active biological organization, founded in 1788. Situated in a grand limestone building next to the Royal Academy, the front hall contained such thrilling relics as a "pencil case alleged to have been used by Linnaeus." Completing the shrinelike feel, a small bronze statue of the naturalist himself stood in the library, where visitors left floral offerings before the buckles of his shoes. When I told the prim librarian that I was interested in learning about Linnaeus's work on fishes, however, her smile inverted. "Linnaeus didn't like anything *slimy*," she said. "Obviously fish weren't his first love."

She explained that Linnaeus had a friend named Peter Artedi, whom he met in 1729, when both were students at Sweden's Uppsala University, outside Stockholm. The two shared a passion for natural history at a time when virtually no curriculum existed in the subject. To avoid competing with one another, they agreed to divvy up the entire living world. While Linnaeus claimed the plants (his first love), birds, and insects, Artedi took all the slimy stuff, including the fish (*his* first love), frogs, and reptiles. The furry critters, not yet called mammals, they shared. The duo made a further pact that if one of them should meet a premature end, the other would see to the publication of his friend's work.

Artedi was two years older than Linnaeus and generally acknowledged to be the greater intellect. Inclined toward deeper analytical reflection and painstaking thoroughness, he often spent several whole days hunched over a single fish, counting and recounting its fins, rays, and vertebrae, and describing every detail of its being: the head, body, scales, eyes, nostrils, tongue, and so on. As he burned the midnight oil, he was laying the foundation for the modern discipline of ichthyology—and perhaps even zoology—including most of the methods still in use today.

In 1735, six and a half years after Linnaeus and Artedi first met, both men were living in Holland, cataloging the private collections of wealthy Dutchmen. On a clear but moonless night in late September, Artedi left a dinner party at his patron's house in Amsterdam and set off on the short walk back to his apartment, a route he knew well. The following morning, his body was discovered floating faceup in a canal, fully clothed in a coat, hat, and wig, with one shoe missing. A contusion on his head suggested he had hit the stone embankment as he fell.

Despite the fanciful thinking of the ichthyologist Theodore Pietsch—whose historical novel *The Curious Death of Peter Artedi* recounts the story above—there's no evidence that Linnaeus was involved in Artedi's demise. But Pietsch is not alone in suspecting that the famed naturalist may well have gotten many of his big ideas from his deceased friend's work. At the time of Artedi's death, he had just completed the manuscript of his masterpiece, *Ichthyologica*. Though Linnaeus kept his promise to publish it, he took two and a half years to do so. During this period, he brought out ten books of his own, securing his reputation as the father of all things taxonomical, even as his work mirrored the classification structure that Artedi had devised.

Living to the age of seventy, Linnaeus spent the rest of his years as a venerated academic, sending students he called "apostles" across the globe to pursue new species. Six of these nineteen young men perished on their quests, including one who was hacked to death in Senegal and another who slit his own throat on the Russian steppes. The very first student to meet a gruesome end succumbed to a tropical fever in Asia while trying to acquire a live goldfish—then a rare and coveted prize among European royalty.

It was a time when people were willing to die for a fish, or at least the larger cause it represented. During the great age of natural discovery, which lasted more than two hundred years from the eighteenth into the twentieth century, the focus of exploration shifted from mapping unknown lands to cataloging the plants and animals within them. Empowered by the do-it-yourself ethos of the Romantic era, naturalists wielding nets, jars, pins, and collecting boxes expanded the number of known animal species almost a hundredfold, from the 4,162 listed by Linnaeus in 1758 to the 415,600 on the books by the late 1800s.

In the aquarium hobby, I had discovered a throwback to this age, perhaps no more so than in the great German species-seeker Heiko Bleher, whom I'd met at Aquarama. Heiko might as well have strolled out of the nineteenth century—the *early* nineteenth century for that matter, before the professionalization of science, when amateur naturalists were the knights-errant of the day.

"I don't think there is any single person in the aquarium world who has such hordes of totally blinded fans just running after him and screaming, 'Heiko! Heiko! Heiko!'" one insider told me at Aquarama.

Nor, however, was Heiko lacking enemies. People seemed to love him or hate him with tremendous passion, sometimes feeling both ways at once. Above all, he exuded the peculiar appeal of an anachronism: the swashbuckling adventurer, the debonair maverick. Not one to be halted by a crocodile breeding ground, he was—I felt certain—a man who could've delivered me to the wild arowana.

Not that I ever expected to see Heiko again. The summer I returned from Borneo, I was eager to put fish out of my mind, though that proved difficult. On the long trip back to New York, my failure to reach the habitat of the Super Red gnawed at my gut like the dysentery I'd picked up, probably from eating that sun-dried fish belly in the swamp. In my determination to find the wild arowana, I'd extended my trip and, as a result, had to take such a circuitous route home that I almost broke a promise I'd made to my mother to be back for a wedding reception—my own.

I'd actually gotten married six months earlier, on an icy December evening in Manhattan, at a restaurant in Union Square, the kind of place with heavy taupe curtains and architectural desserts. It was the same week that I first met Lieutenant Fitzpatrick at his office in Queens, a time when the pursuit of the arowana was not even a thought in my mind. So Jeff (the groom) had no reason to suspect he was marrying a woman with anything other than a totally healthy level of disinterest in pet fish.

Though the wedding itself had been small, with only our immediate families present, the reception I'd traveled back for would be much larger. It was June, and the bulbs had only recently bloomed in the grassy backyard of my parents' suburban stone house in Wilmington, Delaware, where the party was to be held. Meanwhile, I'd cut my return from Borneo so close that a single flight delay would have risked leaving Jeff to fend for himself among my nineteen aunts and uncles, explaining how I'd jilted him for a tropical fish.

That probably goes a long way to explain the relief that swept across his face when I stumbled through our apartment door with a day to spare. A knot inside me unclenched at the sight of the dark-haired man sprawled on the couch, who sprang up, removed a giant poster of prize-winning koi from my hands, and wrapped his arms around me. Then he pulled back, his eyebrows arched in surprise. "There's *less* of you," he said. It was true. I had found an unconventional yet highly effective wedding diet—manic pursuit of the world's most expensive aquarium fish.

I was glad to be home, despite the nagging feeling that I'd left unfinished business in Borneo. Muzzy-headed and spent, I curled up on the couch like a squirrel in a nest twenty-five stories up, chattering semi-coherently about headhunters and how close I'd come to the habitat of the Super Red. Outside the living-room windows, the spire of the Empire State Building winked with reassuring familiarity amid the lights of the city.

The next thing I knew, I woke up in the black night drenched in sweat, my heart drumming, unsure where I was (my bedroom), much less *who* I was. After careful consideration, I concluded the room was a swamp, mistaking the bed for a longboat and the wooden floor for black water. Who the man next to me might be, a lump of covers rising and falling ever so gently, I could not fathom. For a long time, I lay as still as possible. Finally, I reached my hand tentatively over the edge of the bed, toward the gleaming dark surface, and was shocked by the cold, hard wood against my fingertips. It all came back to me then like a drop of ink spreading through water.

"Did anything bad happen to you in the jungle?" my mother asked me the night of the reception, eyeing me with a frown. Actually, she wasn't frowning at me so much as my hair. The previous afternoon, I'd gone to get it cut for the party and spontaneously decided to dye it red. Though it had never been anything but the same ordinary blond all my life, I felt compelled in that moment to express some new aspect of myself. (It's not lost on me now that the swatch I selected was as super red as the fish.) When I looked in the mirror, unable to recognize my own face, I panicked and got the woman to bleach out the dye. Consequently, my hair was now a shade of orange that might aptly be called orangutan. A mother notices such things.

The stress of chasing fish gangsters had finally caught up to me. I hadn't been sleeping well in Borneo, with one earplug in to keep out the roaches, and one ear open (the pillow-facing one) to remain watchful through the night. With good reason, as it turned out, because the trifecta of bogeymen I'd so readily brushed off in my desire to reach my goal—Islamic terrorists, the "fish mafia," and Iban headhunters—would soon take clearer shape. Two weeks after I got home, the al-Qaeda-linked group Jemaah Islamiyah dispatched suicide bombers to the breakfast buffets of a Ritz-Carlton and a Marriott near the hotel where I'd stayed in Jakarta, killing seven people and injuring fifty. In Malaysia, an arowana farmer needed five stitches to the head after thieves ambushed him and stole two hundred of his fish. Perhaps most terrifying of all, I found stomach-churning documentation of the beheadings—even barbecues of human flesh—that Hery had told me about in Borneo.

To answer my mother's question, nothing terrible had happened to me in the jungle. I'd simply failed to find the fish—to connect point Z (that translucent albino arowana) to point A (the wild arowana), drawing a line from the artificial world in which the species ended up back to the natural environment that spawned it. That was all. Yet I didn't feel entirely like the same person. It was as though I had touched my tongue to something not sweet but sour, yet addictively so. Now I felt compelled to take a bite, to keep taking bites until I'd devoured the whole bitter fruit. I felt sure that if I had just reached the Iban, the habitat of the Super Red, the real wilderness, I would have succeeded. The problem was that I'd been so *close* to the wild fish when it slipped from my grasp.

CHAPTER SEVEN

The Explorers

NEW YORK

By the following month, July, the nightmares had begun to abate, and I would probably have finished up my story and moved on from the arowana had I not unexpectedly heard from Heiko. He had just returned from the Australian outback to the Italian countryside where he lived, insofar as he lived anywhere, and was heading to Tehran to give a lecture series on aquarium fishes. Heiko loved Iran. Everyone was especially big on fish there because sharia law forbade the keeping of dogs and frowned upon cats. From the Middle East, he was bound for a fish convention in Los Angeles and had arranged a flight with a five-hour layover in New York. "You invite me for a Italian espresso?" he wrote, and suggested the New York Aquarium. It wasn't the first place I would have thought of getting espresso.

The New York Aquarium is all the way out on Coney Island at the southern tip of Brooklyn, an hour by subway from where I live. To meet Heiko at the airport and then get to the aquarium and back in such a short window, I'd have to drive—something I'd almost never done in the city. The weekend before Heiko's arrival, I rented a Zipcar and did a practice run of the thirty-mile, two-legged trip with Jeff. That Thursday, I rented a second car on my own, cautiously navigated Grand Central Parkway to the short-term parking lot at JFK, and walked over to the international arrivals hall to wait for Heiko.

Eventually, I would grow to be an expert at waiting for Heiko. At the time of this writing, I have waited for Heiko on four different continents. I have waited for him for ten days on the hot, thirsty streets of Yangon, being followed by a suspicious monk. I have waited for Heiko at a pet expo in Nuremberg after he asked me to watch his booth for a moment, saying he'd be right back, and then actually left the building and drove to a *different town*. One time, while I was waiting for Heiko in Bogotá, I felt a weird shaking and rolling sensation, and it turned out I was in an earthquake. I have waited for Heiko outside a train station in the Italian countryside next to a public phone that wouldn't stop ringing as I watched the sun ascend, and then descend, in the sky. At one point, a kindly African woman bought me a chocolate ice cream. To be polite, I had to eat it even though I am somewhat allergic to chocolate. When I got hives on the palms of my hands, it wasn't clear if the hives were a product of the chocolate or of the stress of waiting for Heiko.

Yet as I stood at JFK, bouncing the rental car keys in my hand, it was all still a novelty. Besides, this first time the delay wasn't Heiko's fault. His flight was an hour late, and then there was some holdup with the baggage or customs. In fact, for a good forty minutes or so, he was so close that I could see his black bush hat bobbing impatiently in the crowd on the other side of the glass divide.

Fortunately, I had thought to bring a book—the one Heiko had given me at Aquarama, his mother's memoir. Now I opened it for the first time since Singapore and scrutinized a sultry glamour shot of a blond woman in her forties wearing a black pillbox hat and nuzzling the white feathers that draped theatrically down her cheek.

"This book tells a story true in every detail—that of an adventurous woman, Amanda Flora Hilda Bleher," began Heiko's foreword. He explained that the memoir recounted his mother's explorations of the Amazon in the 1950s, when she took her four children, aged eight (Heiko himself), ten, twelve, and fourteen in search of the much-sought-after discus—then the world's most expensive aquarium fish. The family ventured deep into Mato Grosso (literally "thick forest"), a Brazilian state bigger than France and Great Britain combined. It was the region that had swallowed Colonel Percy Fawcett in 1925 when he went looking for the great civilization believed to have once thrived in the Amazon rain forest. I had just read David Grann's *The Lost City of Z*, about the Fawcett expedition, so I knew, for example, that an estimated one hundred would-be rescuers had subsequently died on more than thirteen expeditions trying to uncover Fawcett's fate. Who in their right mind, I wondered, would take four children to a place where, as Amanda herself put it, "every week people disappeared without a trace . . . where wild Indians lived, where malaria, typhus, and leprosy were rife, and where the rivers teemed with crocodiles, piranhas, poisonous stingrays, and electric eels"?

Born in Frankfurt in 1910, Amanda lost her mother as an infant and was raised by her father, Adolf Kiel, a perfectly round ostrich egg of a man with a black toothbrush mustache, who had entered the aquarium trade in the 1880s. As a young woman, Amanda helped her father run his aquatic-plant nurseries, until at age twenty-eight, she married a butcher named Ludwig Bleher. Four children followed in rapid succession, with Heiko, the youngest, born in a Frankfurt bunker on October 18, 1944, the year the Allied forces flattened the city. By that time, Ludwig was serving in the Nazi army. Eventually, he was captured in France and held in an American prisoner-of-war camp. When he finally came home in 1946, he was a changed man given to fits of rage. Amanda divorced him, getting custody of the children, and opened her own pet business (according to Heiko, "the first mail-order animal business on Earth"), specializing in unusual species. She soon acquired a Nile crocodile for her family, and in 1950 gained a degree of notoriety when a trunk full of eighty venomous snakes was stolen from her parked car in Munich. Learning what was inside, the thief abandoned the trunk on the street, where it was discovered by a group of children, who were then found playing with the snakes. After that incident, the press touted Amanda as "the woman with the heart of Tarzan," writing, "Her life is as colorful and scintillating as her ornamental fishes."

"Ciao," Heiko said, barreling out of customs at last, wearing tight jeans, pointy shoes, and a loose button-down shirt with a leather satchel slung over his shoulder. He kissed me on each cheek, his beard scratchy as a coconut, then announced that his connecting flight had been moved up two hours—he no longer had time for espresso at an airport café much less the New York Aquarium. Waving his hand in a gesture of regret, his eyes caught on the face of his watch, and he indicated that I should follow him at a gallop pace. As we trotted through the airport, he dictated an e-mail for me to send the man picking him up in Los Angeles, letting him know the new arrival time. Another "ciao," and then Heiko was gone, dashing through security. I stood watching his black hat recede with the sensation of having stayed in a revolving door one beat too long, ending up right where I started.

Back in the baking rental car, I turned the key and let the hot air-conditioning blow in my face. Then I opened Amanda's memoir and kept reading.

In 1953, in her early forties, she was drawn westward to the discus like a plant growing toward the sun. Over the next two years, she and her children fell ill with malaria, dysentery, and painful worms that burrowed out of Heiko's scalp. Skirting the edge of believability, the book recounted how the family of five was stranded in a leper colony, chased by a horde of naked Indians wielding blowpipes, and finally survived a deadly bus accident in the Andes, which left Amanda permanently crippled, reliant on a cane. Through it all, they encountered fantastical wildlife deep in the Brazilian rain forest—though, sadly, not the discus they sought. According to Amanda, her youngest child, Heiko, was also the most sensitive. She recalled his crying when he witnessed the lassoing of a puma. "Heiko would have liked to set the puma free," she wrote. "But not until many years later did I truly realize the extent of the puma's suffering."

It was a cryptic reference to the three years she would spend circling a prison cell like a fish in a tank upon her return to Frankfurt in 1955, when she was "branded a spy on the basis of false testimony," having traded fishes in East Germany. With her children left to fend for themselves, Amanda was in solitary confinement when her father, Adolf Kiel, was discovered dead. "Had he fallen down the steps in the boiler-house," Amanda writes, "or had he been pushed?"

When she was finally released in 1958, Amanda was determined to leave Germany forever. Collecting her children, now all teenagers, she departed for Brazil, where she started a fish-and-aquatic-plant exporting company outside Rio. She lived there for more than thirty years before dying in 1991, just shy of her eighty-first birthday.

SIX MONTHS AFTER our whirlwind encounter at the airport, Heiko e-mailed me that he was coming through New York again in March for the annual dinner of the Explorers Club, which was inducting him as a fellow. He invited me to join him for dinner the night after the gala, and I immediately accepted, eager to ask him all the unanswered questions I still had about the arowana.

"You know, I just did my 856th expedition," he told me over sushi in midtown Manhattan, stroking his beard with an expression of supreme self-satisfaction. "At the Explorers Club, no one believed me. Not one of all their members has ever done so many explorations." (Later I did the math: this came to more than one expedition a month *since birth*. Was that even possible?)

"It seems like the explorer is a dying breed," I ventured.

"You're right," Heiko said, nodding gravely, as he poured hot sake into our black lacquer cups. He told me that the theme of this year's Explorers Club dinner had been outer space, and that he wished people would focus on our planet first before worrying about the stars. "So anyhow," he said, holding up his sake. "*Salute!* Nice to see you."

He had just gotten back from the remote region of the Colombian Amazon to which he'd invited me at Aquarama. While there, he said, he had cataloged five hundred fish species, at least 10 percent of which were unknown to science, among them a new discus. I wasn't sure what to make of this.

In fish circles, I'd heard rumors that Heiko wasn't welcome in Brazil since his 2008 arrest for biopiracy—allegedly trying to remove preserved specimens from the country without permission—so now he was entering the Amazon by way of guerrilla country. He told me the Colombian military had made him sign a waiver saying that if he disappeared, no one would go in to retrieve him. "You know, they always tell me, 'It's very dangerous, you don't get out alive.' I never believe these stories," he said. "It was the most peaceful place in the world. We didn't see anyone. Not even Indians live there. Only wild animals and primary rain forest. We camped out every night. It was paradise."

"Are you ever home?" I asked, wondering about the nineteenth-century villa he owned in the Italian countryside near Milan.

"Actually, this year is going to be the first time I'm going to be home more often," he said, swallowing a chunk of raw salmon. "Because my wife's expecting a baby. First one."

Heiko, who had a long-standing reputation as a ladies' man, had gotten married around the same time I did. At sixty-five, he was one year older than my father; at thirty, I was one year younger than his wife. He'd met Natasha in Uzbekistan when she was still a teenager, the university student he hired as his translator. I tried unsuccessfully to picture Heiko as a family man. He didn't even like domesticated animals, declaring them too stupid to tolerate. He insisted that when he was a child, all his pets were wild—like the otter that caught him fish every day in the Amazon until some vengeful Indians allegedly poisoned it, much to his eternal grief.

"I used to have in my bed every night a three-meter boa constrictor," he once told me. "She never shit at night. Very clever animal." Yet her intelligence paled by comparison to that of his capuchin monkey: "I mean, the guy, he understood four languages. I could talk to him, he was always doing the right thing. I gave him a hammer and nails to nail the boards in. I gave him soap to wash my clothes, and he was washing my clothes." This did seem like a helpful pet to have around the house.

Now Heiko, who was home in Italy six weeks out of the year at most, was about to embark on the great experiment of domestication. He figured he should be around when the baby was due that summer. "Expeditions I will only do in the fall, four or five," he said. "Definitely three in Amazonia, and one in Ecuador, and I probably have to do one more in Iran, because they invite me again, and I wanted to do one more in Africa. . . ."

I sat with my chopsticks frozen in the air, letting the long list of destinations wash over me. Who was this man who claimed to be the only person in the world with all four strains of malaria in his blood (*P. falciparum*, *P. vivax*, *P. malariae*, and even the rare *P. ovale*)? I had the sudden urge to reach across the table and poke him to see if he was real. Instead I sketched a map of Borneo on my napkin, starring what I believed was the habitat of the Super Red, and slid it over to him. "Did you get to this lake?"

"Yes," he said breezily. "It was in the eighties. I haven't been back there for twenty years at least."

"Did you see the fish?"

"Yes."

"A big one?"

He shrugged. "You see them. If you stay a few days. You normally see them jumping. It's like the Amazon."

"Maybe not anymore."

"No, yes, sure. It depends where you go."

I tapped my pen thoughtfully on the table and looked through the list of topics I'd wanted to cover: the kidnapped Japanese fish importer (Heiko said he'd disappeared from the fish world); the allegations that the arowana contests were fixed ("As you have probably noticed, it's a real mafia now, but I still don't think so"). Just one topic remained—something I'd heard repeatedly in Southeast Asia: the idea that CITES itself might be responsible for the exorbitant value of the Asian arowana. With all the greed and graft surrounding the fish, it felt absurd to point the finger at a pioneering conservation agreement, protecting some thirty-five thousand species. I thought Heiko might laugh at me or even get angry. I certainly didn't expect his actual response.

His eyes, bleary with jet lag and sake, darkened. "There should be no animal protection at all," he said.

In his view, the quickest way to transform a species into a luxury good, thereby ensuring its extinction, was to declare it officially protected. He told me how as a young boy in Brazil, he awoke every morning to the sight of hundreds of thousands of blue parrots with gray heads rising from the trees to dry their feathers in the sun. There were so many of them they darkened the sky. Now that parrot is extinct in the wild. "But why?" Heiko asked. "Because in 1975, they put it on a list to protect it."

He was referring to the Spix's macaw (*Cyanopsitta spixii*), named for the German naturalist Johann Baptist von Spix, who collected the first specimen in 1819 on a bank of the Rio São Francisco in northeast Brazil before venturing up the Amazon, where he caught the first arowana known to science. In 1975, the parrot was listed alongside the Asian arowana on Appendix I of CITES, that original list of species banned from international trade. When the market for the dragon fish exploded, high-end collectors were racing to get their hands on the Spix as well. By 1990, only one wild bird remained. It was last seen on October 5, 2000, in the Brazilian state of Bahia.

"*Você viu esta ararinha?*" "Have you seen this macaw?" asked a poster distributed throughout the country, showing a picture of a Spix, the dark naked skin around its eyes making it appear masked like Zorro. Today fewer than a hundred of the macaws are known to survive, all in captivity, most in the private collection of a Qatari prince. In Heiko's view, the species had vanished as a direct result of being listed on CITES. Writing about the bird's sad demise in his 2006 book *Bleher's Discus: Volume I*, he concluded, "Anything that is 'protected' disappears in the twinkling of an eye."

This seemed rather extreme—especially when I learned the Spix was already notoriously rare by the 1950s, little remaining of the precarious dry forest the species once inhabited. Yet as much as experts might contest Heiko's memory of massive flocks, he was far from alone in his interpretation of what had happened to the little macaw: even among some conservationists, I'd hear the bird cited as an example of a CITES listing backfiring and driving up demand.

Had a similar fate befallen the Super Red? Had it disappeared from nature too, or had protecting the Asian arowana worked? Despite myself, I felt drawn back to Borneo to find out. By the time Heiko returned to New York the following March for the 2011 annual dinner at the Explorers Club, I had a plan.

I HAD NEVER heard of the Explorers Club before Heiko mentioned it. When I looked it up, I learned the group was founded in New York in 1904—though even then it must have existed largely in the romance of the past. A perusal of its roster through the years reveals the increasingly extreme nature of twentieth-century exploration. The earliest members included Robert Peary and Frederick Cook, both of whom claimed to be first to the north pole (it's possible neither reached it); and Roald Amundsen, first to the south pole. In the 1950s, Sir Edmund Hillary and Tenzing Norgay were first to summit Everest. In the 1960s, Neil Armstrong and Buzz Aldrin walked on the moon, while Jacques Piccard and Don Walsh made the first descent some 35,800 feet to the bottom of the ocean's deepest canyon, the Mariana Trench. Although the pressure at that depth was nearly seventeen thousand pounds per square inch, they were shocked to disturb a bottom-dwelling flatfish as their free-diving bathyscaphe touched down in the mud.

Today the club boasts about three thousand members from sixty nations on seven continents. Every March, some thirteen hundred descend on New York for a black-tie gala at the Grand Ballroom of the Waldorf Astoria hotel. In 2011, at the 107th annual dinner, the outer-space theme had been succeeded by the Mayan Prophecy, the prediction that doomsday was upon us, with the end of the world scheduled for December 21, 2012. I arrived during cocktail hour, made my way up the ornate staircase, and wove through the roistering crowd to the bar, keeping an eye out for Heiko as I went. The drink of the evening was the Zombie, fresh citrus juices with rum and wormwood bitters, in honor of the latest recipient of the prestigious Explorers Club

Medal: Canadian anthropologist Wade Davis. In the 1980s, Davis investigated reports of Haitian zombies—mental slaves who appeared to have risen from the dead—and concluded that voodoo witch doctors were keeping victims in a pharmacologically induced trance for years through the use of a neurotoxin derived from dried puffer fish.

The bartender introduced himself as John and told me he'd been working the event for twenty-eight years. He said he missed the days of exotic hors d'oeuvres. "Scorpions, cock-a-roaches, mealworms," he said. "We used to have martinis with lambs' eyeballs in them instead of olives." I asked what had happened to change that, and he muttered something about PETA's getting involved. "You can kill a cock-a-roach, but you can't eat one," he complained, shaking his head ruefully.

The truth was a bit more complicated. The club had a long-standing tradition of consuming exotic animals—a practice that dates back at least as far as the ancient Romans, who had a particular yen for giraffe, and was revived with great enthusiasm in the nineteenth century as part of the general craze for natural history. In 1904, the menu for the first Explorers Club dinner listed roast polar bear; in 1978, it included elephant stew. Predictably, as the club moved from an era obsessed with dominating nature into one that called for its preservation, dining on rare wildlife became a bit ticklish. In the mid-1980s, Britain's Prince Philip, then president of the World Wildlife Fund, resigned from the Explorers Club in outrage after he learned it had served hippopotamus and lion steaks.

The tradition may have died altogether if not for Gene Rurka, a biologist, farmer, and big-game hunter, who joined the club in the 1990s and took on the role of "exotics chairman," overseeing the preparation of such specialties as marinated beaver, goat testicles, kangaroo leg, calf-eyeball fritters, alligator bites, antelope pastrami, and pigeon pâté. Though the club put the kibosh on serving endangered species, Rurka himself has hunted and eaten

African elephant, which he maintains is delicious. Still, Rurka dutifully embraced serving "sustainable" exotics, which in practice meant a lot of bugs. In 2001, a dozen guests, including Rurka's wife, Marianne, ate underdone tempura tarantulas and had to seek medical treatment when they reacted violently to the hair still present on the spiders' legs. That kind of ordeal, as well as the fact that headlines about maggot mousse invariably overshadowed coverage of the illustrious speakers, had much to do with the controversial decision to forgo the exotics the year I attended.

INSIDE THE GRAND BALLROOM, the crystal chandeliers dimmed. Thunder crashed. Blue lightning flickered. Fog machines belched out clouds of smoke as shamans in loincloths, face paint, and feathered headdresses marched through the crowd, whooping and beating on drums. The faux Mayan apocalypse didn't make it any easier to find Heiko. Finally, I gave up and located my seat at a table near the back, settling in to watch the club's honorary chairman, James Fowler, former host of the TV show *Wild Kingdom*, kick off the festivities with an exotic-animal show. This was another long-standing tradition of the Explorers Club, as well as a recurring headache for the staff of the Waldorf Astoria. A few years back, a snowy owl perched on a chandelier and refused to come down. Another time, camels relieved themselves on an elevator, causing a minor flood. This year all the creatures were indigenous to Mayan country: a woozy black jaguar, a ring-tailed coati, and a nocturnal kinkajou just waking up.

Next to shuffle onstage was a tall, lanky man with a tuxedo hanging off his bony frame as if on a wire hanger. It was eighty-one-year-old Harvard professor emeritus Edward Osborne Wilson, one of the greatest biologists alive, certainly the most prominent naturalist of our time, and the world's leading authority on myrmecology, the study of ants.

As a young boy growing up in south Alabama, Wilson was fishing one day off a dock near the Gulf of Mexico, just across the Florida border, when he caught a spiny pinfish, which flew out of the water and pierced his pupil, blinding him in the right eye. He credits the incident with forcing him to focus on small creatures he could hold in his hand and examine up close. This is how he came to entomology. In more recent decades, however, Wilson's view had broadened to encompass the full diversity of life on earth.

Since the 1980s, he had been sounding the alarm about an extinction crisis of catastrophic proportions. Before the global spread of humanity, extinction rates for animals averaged, very roughly, perhaps one species per million each year. Today, according to Wilson, the earth is hemorrhaging life a hundred to a thousand times more quickly. Without abatement, he says, human activity could result in more than half of all species of plants and animals being extinct or on the brink of extinction by the end of the century.

As the audience at the Waldorf fell silent, Wilson leaned on the lectern and took a moment to dig out a pair of oversize tortoiseshell glasses from his jacket pocket. "One of the sacred rites, if there are any that can be so defined of humanity," he now told the crowd, "is to make the full-out effort of learning about and saving the rest of life on earth." According to the "biophilia hypothesis," which Wilson proposed in 1984, human beings have an "innate tendency to focus on life and lifelike processes"—that is, we are hardwired to find beauty in other living things. "We are in the fullest sense a biological species," he writes, "and will find little ultimate meaning apart from the remainder of life."

Seeking to bolster this theory, one team of biologists measured the physiological changes in people contemplating tropical fish swimming in aquariums and found significant decreases in blood pressure. A second study revealed that dental patients were more relaxed while watching fish. One theory is that we may be attracted to "Heraclitean motion," a pattern that is "always changing yet always staying the same," as in ducks swimming, cows grazing, the interplay of light and shade created by drifting clouds, fire in a fireplace, or waves lapping on a shore. All these examples contrast dramatically with the erratic and sudden motions that typically signal danger—a predator leaping, a snake striking. This is the scientific explanation of what Kenny the Fish told me—that "keeping fish gives you a certain kind of tranquillity."

Such studies support the grander idea that our aesthetic preferences are innate, that they evolved on the African savannas where the genus *Homo* first emerged 2 million years ago. There we developed the "savanna gestalt," which is why—Wilson claims—we're drawn to grasslands dotted with scattered groves of trees. As a consequence, we strive to create savanna-like environments everywhere from cemeteries to suburban shopping malls. In urban spaces, we hang pictures of landscapes and animals that re-create our preferred habitat. This mimicry of biodiversity, however, is an imperfect strategy at best. "Artifacts are incomparably poorer than the life they are designed to mimic," Wilson writes. "They are only a mirror to our thoughts. To dwell on them exclusively is to fold inwardly over and over, losing detail at each translation, shrinking with each cycle, finally merging into the lifeless facade of which they are composed."

If we are attracted to art that replicates our desired communion with other living creatures, haven't aquariums, then, come full circle? Aren't pet fish themselves the most literal representational artworks? The thing itself—captured in a bowl.

"I'VE BEEN LOOKING for you," said a voice, and I lifted my eyes to see a man in a fine Italian suit frowning at me with mild irritation. I recognized the scolding tone, the close-set eyes, the ursine beard, but it took a moment for it all to click. This was Heiko—*without his hat*. I hadn't known he was balding. "Come!" he said, and took off across the ballroom toward a table where a woman sporting large, tinted glasses beneath a gray bob sat wrapped in a floral shawl. It was the legendary Eugenie Clark, the Shark Lady, renowned for her research on sharks and poisonous fishes. Five years earlier, she'd been diagnosed with lung cancer and given six months to live. At eighty-nine, having left five husbands in her churning wake—divorcing the first four and surviving the fifth—she was still scuba diving regularly.

"Genie, I want you to meet a very nice young lady!" Heiko shouted over the din of the ballroom, pushing me forward. "She studies fish also! Arowana!"

When Clark was a small child in the 1920s, her father went swimming in the ocean one day and never returned. Shortly thereafter, her obsession with fish began at the original New York Aquarium in Battery Park. "Last year I went back," she told me wistfully. "The ruins are still there. I saw the remains of the shark tank."

By the 1950s, at a time when female marine biologists were a rarity, Clark was wholly devoted to studying sharks, the sea's most misunderstood gangsters. The release of *Jaws* in 1975 was a boon to her work, propelling sharks into the public eye. But she spent most of her career challenging the misconception that sharks are hateful man-eaters. In fact, while sharks generally eat no more than a dozen people each year, we eat around 26 to 73 million sharks. The only time a shark ever came for Clark, she bopped it on the snout, and it promptly retreated.

"Very, very dear lady," Heiko said, as we stepped back from the table. "Reminds me a lot of my mother." It occurred to me that if Clark had been fixated since childhood on the sea that swallowed her father—dedicating her life to demystifying its monsters—Heiko too was driven by an intense, semiconscious filial piety. He couldn't stop chasing the discus fish, that perfectly round, slimy siren that had lured his mother to the Amazon more than half a century earlier. He had just completed his magnum opus, the thirteen-hundred-page *Bleher's Discus: Volume II*, only to decide that he really ought to produce a *Volume III*.

Heiko's baby daughter had been born the preceding summer, and predictably, he had named her after his mother. Now he told me he'd gotten a text message from his wife. Across the Atlantic Ocean, in a farmhouse outside Milan, Amanda Flora Bleher, at eight months old, had just taken her first steps.

I'D TOLD HEIKO the broad outline of my plan and he had promised to "talk all over" the following day. So early the next morning, I took the subway to the Upper East Side and trudged through the slush of late winter to a grand Tudor-revival mansion, the longtime headquarters of the Explorers Club. Inside, the walls were lined with dark wood paneling. Light filtered in from 114 stained-glass windows. A stuffed polar bear on the second-floor landing emitted a weak mechanical growl as I passed.

I climbed the stairs slowly, taking in oil paintings of African landscapes and bleak arctic scenes. Not until I reached the sixth floor, the very top of the building, did I spot Heiko standing in the soft glow of a skylight in the hall dedicated to honored members. He was wearing a black sweater over a collared shirt and a tie featuring teddy bears that were themselves wearing ties. As usual, he had on his black wide-brimmed hat, which fit in with the headgear in the portraits on the wall he was examining, faces framed by thick fur hoods, pith helmets, aviator goggles, and space suits.

"Theodore Roosevelt," he murmured, as he inspected a black-and-white shot of the twenty-sixth president. After Roosevelt's humiliating defeat in the election of 1912, he took off on an expedition down a black, uncharted tributary of the Amazon called the River of Doubt. Though hardships drove him to the brink of suicide, he still managed to write about the native fauna. Heiko blamed him for giving the piranha a bad reputation.

Most of the wall of portraits was devoid of women, who were not permitted to join the Explorers Club until 1981. I stopped to admire a photograph of Jane Goodall, whom I'd gone to hear speak at New York University just two weeks earlier. Afterward, feeling a bit silly, I waited in a line that coiled through the building to have her sign my dog-eared copy of *In the Shadow of Man*. Despite the assembly-line context, when I got to the front, I nearly choked on the words I'd picked to say: "This is one of my favorite books." ("Favorite book," I'd decided, seemed like too much.)

In a jarring juxtaposition, her Explorers Club portrait hung next to the Trophy Room, where Heiko now wandered. Animal heads plastered the walls, the skins of big cats draped the furniture, and the giant tusk of a woolly mammoth dangled from the rafters. Overlooking the room was a full-length portrait of the Danish explorer Peter Freuchen, who hammered off his own frostbitten toes while mapping Hudson Bay.

"Should we discuss it now?" I asked.

Heiko nodded, stopping before a small circular table next to a three-foot-tall whale penis. I reached into my bag and brought out a map of Borneo, which I unfolded and smoothed out, my hand coming to rest on its blue heart—Sentarum. I explained to Heiko that I wanted to find out once and for all whether the Asian arowana was still there in order to decipher the true impact of commodifying the species. I believed he was the one person who could help me.

"Are you interested in going?" I asked.

I was nervous. The night before, I'd intercepted the anthropologist Wade Davis—that charismatic sandy-haired hunter of Haitian zombies—on his way to the men's room and asked about his work in Borneo. I learned he hadn't been back to the island since his Swiss colleague Bruno Manser mysteriously vanished there in 2000. "He was clearly murdered," Davis told me. When I asked whether Davis had encountered the Iban who controlled the habitat of the Super Red, he said he hadn't, calling that part of Borneo "the end of the Indonesian world." This gave me serious pause: if Wade Davis, an explorer-in-residence of the National Geographic Society and one of the world's most accomplished anthropologists, characterized a place as hard to get to, I realized my own chances were not so hot—not without Heiko at least.

Now, as the morning light filled the Trophy Room, Heiko placed his finger on the map, thoughtfully drawing circles around the lake region. Then his hand drifted clear across the island to the farthest possible river. "The Mahakam," he said. "This is where I have wanted to go for a long time." He told me a unique freshwater dolphin species lived there, the only one in the world he had not yet seen.

"What about exploring Sentarum?" I asked again.

"Well, it's been done extensively by Tyson Roberts, you know," Heiko said dismissively. "He published a book about it."

I did *not* know that. In fact, I had never even heard the name Tyson Roberts. I felt a deep, reflexive pang of disappointment that the swamp had already been explored.

"He's American, but he lives in Thailand. At this moment, he's in Panama. I will see him after the Amazon expedition." Heiko was in transit to Colombia again. After Panama, he was off to Iran, then the Philippines, Singapore, Australia, and New Guinea. He had to be in Tajikistan in August to collect the Heikomobile, a custom-made semiaquatic Mercedes, which had broken down on the Kyrgyzstan border two years ago and was finally repaired. From there, he was waiting on a permit to drive into Tibet, where he hoped to find the highest-altitude fish recorded in history, living above twenty thousand feet. As he recounted these items on his itinerary, I could feel my grand plan for us to find the Super Red together slipping like sand between my fingers. Heiko clearly had no interest in going back to Sentarum and, in any case, no time in his schedule. I'd have to try to contact Tyson Roberts.

The clubhouse was filling up with members now. Heiko propped up his leather satchel and fished around inside. "I brought you something, but they got mashed in the bag," he said, pulling out a squashed box of Italian chocolate-covered cherries, the kind that come with literary quotations. "Read all the messages."

On the subway home, I unwrapped one. A thin piece of tissue paper read in four languages, "The most commonplace of objects becomes charming as soon as you hide it." I picked the chocolate off the cherry and popped it into my mouth. As I chewed, I considered how aptly the message applied to the wild arowana. The next candy contained a line by an Italian poet: "I wanted, I always wanted, O how much I wanted!"

CHAPTER EIGHT

Naming Rights

NEW YORK → SINGAPORE

W hen I finally tracked down Tyson Roberts to a rental apartment in Panama more than a month later, I found him convalescing from a serious attack of deep-vein thrombosis, a blood clot in his leg, which had landed him in the hospital for eight days. He had gone down to Panama as a research associate with the Smithsonian Tropical Research Institute, but now he had to stay off his feet. "I'm more of an embarrassment to them than anything," he said miserably when I called him. "I've got to modify my life totally." He was nearly seventy-one and, for forty years off and on, had made his home base in Bangkok. Now he couldn't fly home in his condition because the pressurization might kill him by dislodging the clot and sending it straight to his lungs, causing a pulmonary embolism. In other words, he was stranded. To make matters worse, he was about to lose his academic housing. He had no idea where he'd go.

I'd looked up Tyson online, but it would take me a good while to realize exactly whom I had called—the grand old man of ichthyology, someone who had probably killed and pickled more species of fish than anyone else alive. Part of my failure to appreciate his renown stemmed from his sparse institutional affiliation. He seemed to be essentially a free agent. Years earlier, Tyson had had a falling-out at Harvard before taking off for bluer waters, much like Aristotle spurned by Plato's Academy. Unlike Aristotle, however, Tyson never ceased his wanderings. Over the course of half a century, he had developed a reputation as an obstreperous loner. "He can be pretty spiny," a young American ichthyologist would later tell me. "All you need to do is cross him once and you are cursed."

As I held the receiver away from my ear, however, all I knew was that Tyson sounded as if he were speaking through a megaphone. And his voice apparently wasn't the only oversize aspect of his person—he complained that he couldn't find compression stockings in XXL. Largest of all seemed to be his Promethean intellect. His mind was like a labyrinth of interconnecting corridors, blind alleys, and hidden trapdoors that was almost impossible to exit once you'd entered.

"Now, do you know the nasty, sneaky, shitty little secret about the dragon fish trade?" Tyson said suddenly, almost twenty minutes into the call, after a series of hairpin detours. "There's a big industry in doctoring the fish." In Thailand, he had heard that Asian arowana were dyed with the same artificial food coloring used in cherry and orange sodas. It made him think of the original margarine of the 1950s, a white, lardlike substance, which had come with packets of orange pigment to mix in.

"Hold on," Tyson said suddenly. "I've got something burning on the stove. I hope it's just water."

It was not. It was franks and beans.

Talking about the market for fake arowanas got Tyson onto the subject of art history, specifically ancient Cambodian statuary, on which he was a considerable expert. His favorite works—in his opinion, "the greatest art ever produced"—were what he called the "portrait statues" of the Khmer kings dating back to the fifth century. He estimated that only about one in twenty of those found in museums was authentic, and it had become a hobby of his to expose the fakes.

Getting Tyson to talk about Borneo proved to be a bit like corralling a ferret. But I eventually managed to piece together the bones of the story: Back in 1976, shortly after leaving Harvard, Tyson had made an expedition to the island on behalf of the Smithsonian Institution. He was drawn to the Kapuas River, the longest in Indonesia, because it was probably the richest in fish fauna, having once been part of the vast Sunda basin, which rivaled the Amazon and the Congo in size. Tyson embarked on the journey with an Indonesian scientist who was summoned back to Jakarta shortly after they set off up the Kapuas. The Indonesians expected Tyson to turn back then too, but he didn't, instead continuing on with a crew that didn't speak English. To the vexation of the officials back in Jakarta, who were nervous that an American might go AWOL on their watch, Tyson spent another six weeks collecting fish in Sentarum, which he found to be relatively wild and untouched.

"I'm sure the Kapuas today is largely ruined," he said. "My feeling is this: if I go to a place, within ten or thirty years after my visit, there won't be any fish there anymore. That's just about true of anywhere I've ever been my entire life."

"Did you actually see the Asian arowana?" I asked.

"Oh, I collected some of them."

He didn't remember where he had caught the species exactly—he hadn't opened his own book *The Freshwater Fishes of Western Borneo* in years—but the fish had been plentiful during his research in the mid-1970s. He had a vague memory that the specimens had been colorful though not spectacular. He did clearly recall that not long before his trip, the fish had been recognized as an endangered species first by the IUCN (the International Union for Conservation of Nature) and then by CITES. "In my opinion their information was totally inadequate," he said, adding that the source for such listings was often the head of a fisheries department who hadn't been in the field for years. Like Heiko, Tyson felt that designating an animal endangered was not always a good idea: "You just push the market value up, sometimes like hell."

I told Tyson that I wanted to retrace his steps and find out what had become of the Super Red arowana in the intervening years, determining whether it still existed in the wild. "I've been trying to get Heiko to take me," I mentioned.

"If you do go with him, you better be prepared to take care of yourself," Tyson warned. "Because he might leave you behind in Lake Sentarum."

When Tyson was laid up in the hospital in Panama, Heiko had flown in from the Amazon to visit him, as he'd told me he planned to do, and brought along a get-well present—a tiny new eel for Tyson to describe. But Tyson complained that Heiko hadn't properly recorded the name of the river basin it came from. "He says he's never mixed up," Tyson griped of his friend. "He says, 'I have seen ten thousand species, and I can recognize every one that I have ever seen.' Nobody can do that! God couldn't do that! God wouldn't have *bothered*."

Tyson had been talking a full three hours when he happened to mention offhand a rumor he had heard, something buried deep in his prodigious mind like a banknote misplaced in an office overflowing with seven decades' worth of papers. The rumor was this: a new, as-yet-undescribed species of arowana lay hidden in southern Myanmar, a region long isolated from the outside world by the iron fist of a military junta. He had seen an unconfirmed photograph of the fish, which featured the telltale large scales and dragonlike barbels of an arowana. "But the entire body is covered with a kind of *handwriting*," he told me, "or something that makes me want to name the species *Scleropages scriba*. Like it's been written on. It's gorgeous."

"Are you going to name it?" I asked, picturing the creature

scrawled upon like an ancient scroll and for a crazy moment wondering what it might say.

"I haven't got specimens!" Tyson roared. Then he paused for the first time in hours. An idea was dawning on him. "Why don't you send your husband there and get specimens for me, and then I'll name it after you?"

I laughed. I didn't know whether to be amused that he was suggesting retrieving an apparently irretrievable fish, or offended that he had skipped right over the idea of sending *me*. The only reason Tyson even knew I was married was that several hours earlier, when he was thundering about a problem with his e-mail, I tried to help him and mentioned in passing that my husband was an engineer at Google. But that was all he knew. He didn't know, for example, that Jeff had grown up in the heart of Manhattan, descended from a long line of New Yorkers so thoroughly adapted to city life that my mother-in-law once told me that trees depress her. When Jeff's older brother was a toddler, he allegedly saw a patch of grass and called the mysterious substance "green water." Jeff was nine when his father got a job in rural New Jersey and moved the family there. When they pulled up to the five-bedroom house on a leafy cul-de-sac, Jeff's younger brother asked where the doorman was. The first night, all three boys dragged their mattresses into the smallest bedroom. A month later, they were back in New York. All this flashed through my mind, mixed with the sting of injured pride, when I said reflexively to Tyson, "I think *I'd* be the more likely one to go."

"Is that right?" Tyson shot back with reptilian quickness. "Why don't you do it then?"

"I don't think I'm equipped," I said.

"You *are* equipped!" he shouted. "For goodness' sake!"

"I don't know. . . . We live in Manhattan. . . ."

But Tyson was no longer listening to me. "And a *lady*!" he muttered. "You'll get away with murder!"

The idea of Myanmar (or Burma, as it's long been called in the West) conjured for me flickering, grainy footage of monks marching in saffron robes. I had only a vague notion that the country did not have a warm and fuzzy government. Not until later, when I opened the book *Burma/Myanmar: What Everyone Needs to Know*, did I read the first line: "Myanmar is, after North Korea, probably the most obscure and obscured state in the contemporary world."

"It's almost like being in a past century," Tyson told me, recalling how years ago he had snuck over the Thai border into the rebel-controlled hills and hired insurgents armed with Kalashnikovs to take him looking for fish in the Great Tenasserim River. He remembered what a hell of a time the boys had, diving in the water in the sunshine, as old British bombers belonging to the Tatmadaw, the Burmese army, passed overhead.

"Did you feel safe?" I asked.

"I felt safe every place I've been," Tyson said, "except sitting here right now. I could die at any minute."

Because he had explored the upper stretch of the Great Tenasserim River himself, he knew the arowana couldn't be there. That left the lower half, which he had never managed to get to. He gave me instructions to sneak into an off-limits coastal village with a name he couldn't remember and wait around long enough to gain the trust of the locals. "If you tell them what you're after and they believe you, then that will be enough. If they don't believe you . . . then I don't know, huh?"

"I'll consider it," I said finally, if only to placate him. Although intrigued, I'd spent so long thinking about the Super Red that I couldn't immediately switch gears. Insofar as the key to life, biologically speaking, is flexibility—the capacity to adapt, to improvise, to change one's strategy in medias res— I have no Darwinian advantage whatsoever. I am the dog that stands at the base of the tree the squirrel went up. I will stand there until dark, till I'm dragged inside, then dash into the living room, press my wet nose against the glass, and bark at the tree, keeping everyone up all night. Case in point: For two years, I'd been looking out my office window, gazing across chimney tops and water towers, over the East River and beyond Brooklyn, across the Atlantic Ocean and the Sahara Desert, the Arabian Peninsula, the Indian Ocean, and the South China Sea, to the heart of Borneo. If Heiko wasn't going to take me to Sentarum, I knew I would have to get there some other way. I wasn't about to change my goal now. Any thoughts of going after some rumored fish in Myanmar would have to wait.

I kept trying to steer Tyson back to the subject of Borneo, to Sentarum, to the Iban (he apparently hadn't noticed them, since they were not fish), and of course to the Super Red. But I couldn't get much more out of him. At this point, Tyson and I had been on the phone for more than four hours. I think it was the longest phone conversation I've ever had with anybody. I pictured him alone and infirm, about to get the boot from a mildewy tropical apartment smelling of burnt franks and beans. "Listen," I said, "you take care of yourself."

"Ah, shit," Tyson muttered through the receiver. "I'm not enthusiastic about my chances. I'm pretty happy to talk about the past, because I think it's all I got." But no sooner had he uttered these words than he started in on the Myanmar arowana again, that one last fish that would inscribe his name in the annals of history. He assured me he would survive long enough to describe it. In fact, he could guarantee that he would. It was a fish worth living for.

OVER THE PAST 250 years or so, since Linnaeus invented modern taxonomy, nearly 2 million species have been named and described. Scientists don't agree how many remain to be discovered, but most estimates predict a grand total three times as high, or more. In the words of E. O. Wilson, all we know for sure is that "we have only begun to explore life on Earth."

How hard is it to discover a new species in the twenty-first century? Well, it depends what you're looking for. Of the nearly 2 million species currently on the books, about 1.5 million are animals (as opposed to plants, or fungi, bacteria, viruses, and the like, which remain very poorly known). Of the animals already inventoried, the vast majority—about two-thirds—are insects. Many, many more insects remain to be discovered than have already been described—an estimated five times as many. So finding a new bug is relatively easy.

But if you want to discover a vertebrate with a backbone like your own, your best bet is almost certainly to go after a fish. As Melanie Stiassny, curator of ichthyology at the American Museum of Natural History, puts it, "There are an awful lot of fishes on the planet. We know a little about some of them, but we know virtually nothing about most of them." Amazingly, the world contains as many fish as it does all other vertebrates combined. In other words, you can add together all the mammals, birds, reptiles, and amphibians—every last sparrow and salamander, crocodile and naked mole rat—and that number is barely as great as the total number of fishes. Some 30,000 fish species are currently on the books, with about 350 new ones described each year.

Some of these new fish represent truly novel discoveries. In other cases, fish already known to science are split into multiple species based on new observations. In 2003, this happened to the Asian arowana when a paper appeared in a French journal arguing that the fish was not a single species at all, but rather four. The study contended that the different color varieties—the green, silver, golden, and red—differed from one another, both genetically and in appearance, enough to constitute distinct species. The lead author was a young French ichthyologist named Laurent Pouyaud, and the subsequent backlash demonstrates just how controversial naming a new arowana can be.

Pouyaud's paper caused an uproar in part because how a species is defined can have a tremendous effect on whether an endangered population receives protection. In splitting the Asian arowana into four species, Pouyaud decided the green should retain the name *Scleropages formosus*. He gave new names to the other three color varieties—including the coveted Super Red—theoretically stripping them of their protected status, since the newly named fish no longer appeared on any official lists of endangered species.

For the most part, scientists rejected the study and continued to view the Asian arowana as a single species. Their reasoning, however, probably had less to do with the scientific merit of the argument, or the potential consequences for the wild fish, than with a previous scandal involving Pouyaud and the naming of yet another high-profile creature: the coelacanth (pronounced *seel-uh-kanth*).

A primitive order of fish prevalent in the fossil record, coelacanths were believed to have gone extinct with the dinosaurs until 1938, when a South African woman spotted one in the bycatch of a deep-sea trawler. The five-foot-long, blue-green fish with limblike fins and a "little puppy dog tail" was one of the most remarkable zoological discoveries of the twentieth century. Not only was the coelacanth a "living fossil" (to use Darwin's term), a hanger-on, one of the "remnants of a once preponderant order," but it was also billed as a "missing link" between fishes and land animals. While that's not exactly accurate, it's true that the coelacanth is a rare "lobe-finned fish," with fins that sprout from stubby buds and move in tandem much like legs. As such, it's related to the first fish that walked out of water 300-some million years ago, giving rise to all land vertebrates, including us. The only other lobe-finned fish known to have survived to the present day are the freshwater lungfishes. For a long time, a debate raged over whether the lungfish or coelacanth represented our closest living fish relative.*

So in 1999, four years before Pouyaud's foray into arowana, when he published a paper describing an entirely new species of coelacanth living on the opposite side of the Indian Ocean, this was big news. However, Pouyaud himself had not discovered the fish. Rather, an American ichthyologist named Mark Erdmann had spotted it at a market while honeymooning on the Indonesian island of Sulawesi. Much to Erdmann's chagrin, he let that specimen get away. But he spent the better part of a year searching for another—until at long last he succeeded. While a team of American biologists analyzed the coelacanth's DNA, Pouyaud got access to the specimen at the Indonesian institute where it was stored and, together with five Indonesian scientists, described and named it, failing to mention its discoverer at all.

A swirl of accusations of scientific piracy followed. In response, Pouyaud and two French colleagues submitted a photograph to the prominent journal *Nature*, purporting to show the Indonesian coelacanth at a local fish market in 1995. The point was to prove that the French had known about the fish long before Erdmann did. Just before the article went to press, however, a *Nature* staffer discovered the photograph was a fake—the very shot Erdmann had taken of his hard-won coelacanth photoshopped onto a wooden slab. Instead of the submission, *Nature* ran an exposé. Ever since, Pouyaud had carried the stain of the scandal, and thus his claims about the Asian arowana were largely disregarded.

SO THE ONE attempt to split the species had not gone over well, and no one had touched the subject since. Yet the fish

---

* This was finally settled in 2013 when the Broad Institute of MIT and Harvard sequenced the coelacanth genome, which was then analyzed by an international consortium of experts who concluded that the lungfish wins.

that Tyson Roberts had told me about wasn't just another color variety—it had a pattern, which seemed to indicate that it might be something truly unique. Shortly after our conversation, I found a grainy photo circulating on the Web of an arowana with strange, calligraphic markings for which the fish had been nicknamed the batik. No one knew where the picture had come from. Some thought it was a fake.

However, if an unknown species of arowana still existed in the world, it made sense that it might be in Myanmar, which perches like a monkey between China and India, its long geographical tail dangling between the inky waters of the Andaman Sea and the emerald mountains of Thailand. Much of the country had been effectively sealed off from the world since a military coup in 1962. As an aquarist friend put it, "One day you can go there, the next day you can't come out."

But even if the fish *did* exist, a scientist somewhere could already have a specimen and be preparing to describe it. This possibility weighed heavily on Tyson, who had urged me to do some reconnaissance and ask around "very gently" about the potential new species. "But then keep it very quiet," he warned. "Don't tell the Singapore people, because they're going to want this fish very badly."

I didn't have to look far. Just a few weeks after I spoke with Tyson, I traveled back to Singapore to attend Aquarama again. Over lunch with the molecular biologist László Orbán of Singapore's Temasek Life Sciences Laboratory, I learned that he too had heard of the Myanmar arowana—in fact, he had been sent specimens for genetic analysis.

My heart sank like a stone in deep water. Though I'd set out to help Tyson, the idea of discovering a new species had begun to hook me too.

"That fish is very, very different from the rest—even more beautiful than the ones you find here," said Orbán, a gentle Hungarian with such a deep, soulful love of fish that his eyes moistened when discussing them. "It has a very peculiar color pattern on the scales, really very nice. I only saw a couple of pictures but . . ."

Wait. Only *pictures*?

It turned out Orbán didn't have specimens of the fish itself, but rather two little clippings of its fins. He said that was all he needed for genetic analysis: "Probably less than the dirt under your fingernail." The fin clips had been sent to him by a German ichthyologist named Ralf Britz at the London Natural History Museum.

"I assume that *he* has the fish?" I asked.

"I don't know." Orbán shrugged. "You need to ask him."

WHEN I GOT back home, I called Ralf Britz at the London Natural History Museum. He answered the phone sounding pleasant enough, asked if I could call back in forty-five minutes, then promptly went home for the day, leaving me to spend the better part of the morning dialing his machine at regular intervals. While this didn't exactly endear him to me, I did not yet consider what it might mean. Instead, I reasoned glumly that where there are fins, there is usually a fish. Besides, in this day and age, wasn't a measly fin clip—a little chunk of meat—all a scientist needed to describe a new species? Didn't it contain the entire story of the fish writ in DNA?

Only after more than two months and half a dozen e-mails did I finally get Britz on the phone in London. He'd been avoiding me *because he did not have the fish*. He'd procured the fin clips that he'd sent to Orbán from a source in Myanmar whom he declined to name. While he was keen to describe the new arowana, getting his hands on a full specimen—the actual fish—was, he insisted, "absolutely necessary."

It was also virtually impossible. "Myanmar is a pretty difficult country to work in, as you probably know," he told me in his cool German accent. "The Tenasserim River, where the fish occurs, is up in the mountains, and it's really unlikely that you'll be able to go there."

He was speaking from experience. He'd been to Myanmar seven times in the past decade and still hadn't managed to get to the arowana. The previous year, when he thought he could finally travel to its habitat, the government revoked his permission at the last minute. "I've never seen a country that is so controlled," he said. "They know every step I make."

In retrospect, this did not sound promising. But his pessimism scarcely registered as my heart thrilled to realize the fish remained viable quarry up for grabs. "Do you think it could be a different species?" I asked.

"I can't say anything," he said, picking his words carefully. "It could be. It could not be."

I wasn't sure what rules of etiquette might apply when two ichthyologists were after the same fish, but I entertained a vague, Pollyannaish notion of happy collaboration, everyone working together for the advancement of science. Toward this end, I asked Britz, delicately, if he knew Tyson Roberts.

"If you find an ichthyologist who does *not* know him, then I would be very surprised. He is a very illustrious character. Everyone knows him." He seemed to want to say more, but stopped there.

"Do you know Heiko Bleher?" I asked.

Britz laughed uncomfortably. "You've got them all lined up! Yes, I know Heiko, yes." Same censored silence.

"I actually was hoping that he might be able to come with me to get the fish," I confessed. "Because I think if anyone could—"

Britz groaned and uttered the word *no*.

"You don't think he could do it?"

"Maybe he could pull it off, but I don't think he would be able to legally," he said. "I wouldn't associate with him if I were you."

TOO LATE.

I'd already told Heiko about the inscribed Myanmar arowana when I'd met up with him at Aquarama. He had been drinking espresso, bush hat tilted over his eyes, shirt billowing open to his solar plexus, when I casually tossed out the morsel, then watched as he almost imperceptibly sensed the bait. A slight flare of the nostrils. A subtle shift of the gaze. He remained cool as a sea cucumber. But when I showed him the grainy photo I'd found online, he leaned in close to scrutinize the fish. "That *is* different," he said.

I believed Heiko was ideally suited to go after the rumored arowana for two reasons: (1) He could supposedly get anywhere. Even Ralf Britz conceded, "He's been to areas I fully admit that no one else would be able to go, just because of his persistence." The Tenasserim—the extreme south of Myanmar, one of the last great unexplored frontiers on earth—would be a true test of his abilities. And (2), he wouldn't want to keep the fish for himself. As a field naturalist in the Victorian model, he didn't describe new species, but rather passed on specimens to lab scientists, as he'd done with the little eel he'd brought Tyson as a get-well present when he visited him in the hospital in Panama. Who better than Heiko to fetch the great ichthyologist his *poisson de résistance*?

When I saw Heiko, however, I hadn't yet resolved the source of the fin clips; and I knew he wouldn't want to go after the fish if someone already had it. Plus there were other hurdles to be considered.

"CITES throws in an obstacle," Tan Heok Hui, the "fish man" at the National University of Singapore, pointed out when I caught up with him. "It's supposed to regulate international trade, but it hinders research." He told me that it's such a pain to work with animals and plants listed on CITES that many scientists avoid them altogether. That means that some of the rarest species in the world, which should be getting scientific attention, aren't.

Tan also knew about the batik arowana, but claimed he had no interest in pursuing it—too much of a bureaucratic nightmare. When I confided that Tyson had suggested I go get the fish, Tan said: "No, no, don't do that."

"He seemed to think it was the easiest thing in the world," I said.

"He makes it sound easy. But if you get caught bringing the material to him, you are the mule."

Tyson's perspective, however, was that the fish wasn't protected—how could it be when it was a species brand-new to science? Of course there was a considerable catch-22: no one could say for sure that the fish was a new species, distinct from the Asian arowana, until it had been formally described; yet no one could describe it without first getting their hands on a specimen.

What's more, not just CITES posed a potential roadblock. According to the Convention on Biological Diversity, a global treaty from the early nineties, "states have sovereign rights over their own biological resources." In other words, Myanmar owns its indigenous fish down to the last scale. Going into the country and removing so much as a fin clip without permission may violate national laws and constitute stealing DNA, aka biopiracy. That's how Heiko landed in jail in Amazonia in 2008—for attempting to exit Brazil with preserved fishes in tow.

I figured, however, I could worry about permits later. The first step was to discover if the Myanmar arowana even existed. Once I confirmed that Ralf Britz did not have a specimen in London, I e-mailed Heiko to ask flat out if he wanted to go after the fish with me, noting soberly, "I think there is a serious scientific contribution to be made here." But did I? Was this really about science?

FROM THE BEGINNING, the glory of discovering and naming fish has been a big part of the aquarium hobby. By the 1930s, new species were entering the trade by the scores. In 1935, the leading American aquarium guide listed some 243 species. In 1955, when the godfather of the tropical-fish industry, Herbert Axelrod, published his *Handbook of Tropical Aquarium Fishes*, that number leaped to 410. The introductory chapters focused on how to collect fishes from the wild, how to send an unidentified fish to an ichthyologist for identification, and how scientific names are assigned.

An eccentric and ambitious character from blue-collar Bayonne, New Jersey, Axelrod founded *Tropical Fish Hobbyist Magazine* in 1952, at the age of twenty-five, and grew the business into TFH Publications (dubbed "the General Motors of the pet world"), publisher of such titles as *Modern American Mouse* and *Guinea Pigs for Those Who Care*. He strode onto the scene in the decade following World War II, just as pets were becoming full-blown commodities—the era when purebred dogs became popular and puppy mills were born. Tropical fish were no exception. Once caught from the wild, then bred on a small scale by hobbyists, they were now produced on commercial farms in Florida. Axelrod soon became the largest such breeder in the world, reportedly stocking some 6 million fish on five farms near Tampa. At one of those farms, he hired a young German upstart fresh from Brazil—a cocksure prodigy named Heiko Bleher.

The original legend of the fish world, Axelrod cast himself as an intrepid explorer. In 1965, *Sports Illustrated* reported that he had made more than forty expeditions to South America, Africa, Australia, the Fijis, Indonesia, Thailand, India, and the Malay Archipelago. By his own account, he could identify seven thousand species of fish on sight.

It was Axelrod who launched Heiko to fame as a correspondent for *Tropical Fish Hobbyist Magazine*. For a while, the two were like father and son. Then in 1976, they set out to drive the new Trans-Amazonian Highway together—it didn't end well. (Heiko claims Axelrod abandoned him in the jungle without rations.) After that, they didn't speak for two years. By the late 1980s, however, they had reconciled and begun work on an aquarium book to end all aquarium books, a comprehensive guide to every fish in the trade. But the collaboration soured, and Axelrod, who (according to *Sports Illustrated*) had long been "tremendously fond of quarrels and litigation," sued his protégé. A lawyer who knew both men well described Axelrod's vendetta as a matter of pride: "Herb got into this 'I'm better than you are, and I've got more fish named after me than you,' and he went out to destroy Heiko."

Ultimately, Axelrod won. According to Heiko, he lost thirty years' worth of savings and all seven of the houses he owned at the time. But his deepest regret is that when his mother died in 1991, he had not seen her for two whole years—he'd been too tied up with lawsuits in New Jersey.

Meanwhile, Axelrod amassed such a fortune that he was able to buy an entire quartet of Stradivari stringed instruments, valued at $50 million when he donated them to the Smithsonian Institution in 1998. The timing seemed ideal to get a hefty tax write-off on the sale of his company TFH for $80 million. Soon, however, he was embroiled in a lawsuit with the buyer, who found evidence of cooked books, alleging that Axelrod had used the company to support personal indulgences ranging from cigars to mistresses. This led the IRS to investigate. In April 2004, Axelrod, then seventy-six, was indicted on federal charges of tax evasion and fled to Cuba. Two months later, he was apprehended in Berlin and eventually extradited to New Jersey, where he ultimately served sixteen months in prison.

His downfall notwithstanding, Axelrod's name is immortalized in sixteen species of fish (including an extinct coelacanth) and one entire genus. Most famous of all is the cardinal tetra, *Cheirodon axelrodi*, the bestselling wild aquarium fish of all time. Axelrod claimed to have discovered the small creature with a red belly and neon stripe in a tributary of the Amazon in 1954. But the International Commission on Zoological Nomenclature later found that he had bought it at a pet shop in New Jersey.

Heiko, of course, is the namesake of the rummy-nose tetra, *Hemigrammus bleheri*—which he says is the number *two* bestselling fish of all time. I hoped he wouldn't be able to resist the chance to add another notch to his net.

IT TOOK HEIKO more than a month to get back to me about Myanmar. When his e-mail arrived, on the last day of August, he had just been in a horrific crash, reminiscent of the one he'd survived with his mother in the Andes fifty-seven years earlier. "I am very sorry for this delay in answering you," he wrote, "but I just returned last night from a tremendous expedition across Tajikistan, Uzbekistan, Turkmenistan, and Iran, where I finally after amazing discoveries and disappointments, had a terrible accident on this past Sunday near the Turkish border." (NB: I've cleaned up the typos. The e-mail looked as if it were typed with his toes.)

Heiko had gone to Tajikistan to pick up the Heikomobile, which had broken down there two years earlier and was finally fixed. Tajikistan, however, may not be the most reliable place to get a custom-made semiaquatic Mercedes repaired. As Heiko was cruising out of the Zagros Mountains of western Iran, the entire rear wheel—axle and all—flew off, and he was "blown out of the window," smashing his left leg, which was now in a cast. He had "terrible scarce [*sic*] all over," and "the head was stiched [*sic*] many times."

This was all quite dramatic. Nonetheless I found myself skimming through the grisly details like a prospector sifting through sand until my eyes seized greedily upon a nugget glinting toward the end: "Yes, I would very much like to go back to Myanmar." In particular, Heiko was keen to explore the *aqua incognita* of the extreme south—that long monkey tail, the Tenasserim.

The only problem, he said, was that he had four expeditions to make before December (in a *cast?*), "and I do not even know how many lectures and seminars." But he thought he could work out a trip in February, five months away. That would be dry season in Myanmar, which he preferred for collecting, since more fish are then concentrated in less water. "Now I will sleep," he wrote. "But I think we can do it."

I wasn't so sure. Tyson had made it sound as if collecting the arowana would be easy. However, I'd come to discover that the Tenasserim River, where the fish was supposed to be, was the site of the longest-running civil war on earth.

It was now late summer 2011. The previous fall, Myanmar had held its first presidential election in twenty years, which the United Nations dismissed as fraudulent, and President Obama denounced as "neither free nor fair." After the election, violence erupted once again in the border regions of the country. As I read reports of escalating armed clashes between government forces and separatist hill tribes, I found myself mired in acronyms. There was the DKBA (Democratic Karen Buddhist Army), KNLA (Karen National Liberation Army), the KIO (Kachin Independence Organization), and KIA (Kachin Independence Army), plus Communist rebels and a mujahideen Islamic insurgency, all fighting against, at various times and in various alliances, the Tatmadaw—the Myanmar armed forces.

What's more, I had another concern: that I might not be able to get into the country at all. Myanmar was known to routinely deny visas to journalists and writers, even those traveling as tourists. I worried that one Google search of my name would reveal my professional background and I'd be rejected—leaving Heiko to retrieve the arowana without me.

So that September, I found myself standing in line at the Civil Court of the City of New York, taking Jeff's last name—Korn—so I could get into Myanmar as a tourist without inciting suspicion. The one problem, I soon learned, was that I was legally required to publish the name change along with my address, which defeated the purpose of anonymity.

The only way around it was for me to petition the court to seal the record, an option designed pretty much exclusively for victims of domestic violence. As I sat on the hard courtroom bench, waiting for my sealed name change to be granted, I thought about how unusually nice everyone in the courthouse had been to me, speaking in low, soothing tones, though none would look me in the eye, instead fixing their gaze politely on a point past my ear. That, if I had to pick a moment, was when I began to suspect that my relationship with the arowana was not 100 percent healthy.

IN THE END, I was granted a four-week visa to visit Myanmar in February, when I planned to meet Heiko and try to find the rumored new species of arowana for Tyson. Before then, however, I had some unfinished business to take care of on my own. Rainy season was just getting started in Borneo. According to local fishermen, this was the time of year when it was still possible to catch the arowana. I was determined to learn, at long last, what had become of the Super Red.

PART III

THE
SUPER RED

CHAPTER NINE

In the Age of Aquariums

MALAYSIAN BORNEO—SARAWAK

In 1852, the Austrian explorer Ida Pfeiffer traveled in a wooden longboat up Borneo's River of Man-Eating Crocodiles clad in full Victorian dress. Taking shelter for the night in a Dayak longhouse, she was given the place of honor beside the hearth, where she bedded down beneath two skulls and a freshly lopped-off head. "The wind rushing through the hut rattled the dry skulls continually one against another," she later wrote, "and the vapor and stench from the fresh head was suffocating, and from time to time driven by the wind right into my face. . . . Sleep was impossible, and I got by degrees into a perfect fever of terror."

I first came across Ida's name in a survey of the fishes of Sentarum, just as I was preparing to go back to Borneo on my own, more than two years after my first trip there. "Exploration (and knowledge) of the fish fauna," I read, "started with Ida Pfeiffer, who in [1852] discovered the Kapuas Lake district." Wait—*what?* A European lady in the 1850s, when women were largely confined to the domestic sphere, made it all the way to the swamp I'd found so hard to reach in the twenty-first century?

Ida, I soon learned, had been a popular Austrian authoress famous for chronicling her travels. As a little girl in Vienna, she dreamed that a passing carriage might whisk her away to some far-flung locale. Before she managed to go anywhere, however, she married and raised two children, only to find herself a middle-aged, empty-nest widow teaching piano lessons in her front parlor. Under the pretext of piety, she arranged a trip to the Holy Land, publishing an account of the pilgrimage that sold well enough to fund a second expedition to Iceland. After that, she announced to her two grown sons that she wished to venture to Brazil, failing to mention one minor detail—that she planned to come home the long way around, circumnavigating the globe.

In no time, I had my nose buried in Ida's 1850 bestseller *A Lady's Journey Round the World*. She was not exactly—how shall I put this?—a *nice* woman. She reveled in describing how ugly people were and complained a great deal about inferior help (like the servant with elephantiasis who put a damper on her appetite). The irony was that Ida, finally free to see the world, found nearly all of it wanting. She deemed Rio de Janeiro filthy and the Indians of Amazonia primitive savages. She felt scandalized by the free love of the Tahitians. When she arrived in Canton, where she dressed as a man to roam the streets freely, she found she didn't like the Chinese much either, pronouncing them dastardly and false. After months spent castigating "heathen and infidel countries" across the Near East, she breathed a sigh of relief to finally reach Christian Russia— where she was promptly arrested on suspicion of espionage and thrown into jail. She later wrote that Russians have stupid, coarse features, and their behavior corresponds completely to their appearance. Her book was a hit.

So much so that within a year of publication she took off to research its sequel, *A Lady's Second Journey Round the World*. Departing London aboard a cargo ship bound for the Cape of

Good Hope, she intended to circle the globe in the opposite direction this time, traveling eastward from Europe rather than westward. This journey would ultimately land Ida in the East Indies—modern-day Indonesia—where she became the first European to embed herself among the Batak cannibals of Sumatra, whom she famously told she was too old and tough to make good eating. On the island of Borneo, she landed in the northwest state of Sarawak, controlled by the "White Rajah" James Brooke, an eccentric merchant-adventurer who had claimed the territory for Britain despite having no official commission from the crown. Brooke ruled the coast like a personal fiefdom, while undertaking a violent campaign to stamp out piracy and head-hunting in the interior. Because he'd just been summoned back to London to answer for atrocities against the natives, Ida was disappointed to miss him. Instead she was received by his nephew Captain John Brooke, who attempted to dissuade her from ascending the Batang Lupar—otherwise known as the River of Man-Eating Crocodiles.

Ida, however, was not to be deterred. In January 1852, she set off upriver into the wild territory of the Dayak headhunters. Boggy and miasmal, the vast tracts of primeval forest reminded her of the Amazon, only wetter. The dense jungles shrouded in mist hid pygmy elephants, hairy rhinoceroses, mouse-deer the size of rabbits, rust-red orangutans, bearded pigs, and potbellied proboscis monkeys with noses like bicycle horns begging to be honked.

Aside from the oarsmen rowing the longboat, her only companion was a Malay guide from the coast, whom she quickly wrote off as lazy and impudent. Each evening, as darkness fell, she clambered up another ladder—a petite figure, slightly stooped, petticoats hiked above her hips—to appear in the doorway of the nearest longhouse. To the Dayak, bedecked in necklaces of glass beads, tiny shells, and human teeth, their forearms and calves encircled in copper rings, their heads adorned with lofty plumes plucked from the speckled argus bird, the visitor in her full skirts must have seemed a bizarre vision. She was pale with stern black eyes and thin lips set in a straight line of opprobrium. To shelter herself from the equatorial sun and torrential rain, she wore on her head not only a bamboo hat from Bali but also a giant banana leaf.

Seventeen days after the start of her journey inland, Ida reached the base of a mountain chain, which she crossed barefoot, shedding shoes and stockings to muck through an endless succession of streams and morasses. Vines wound round her legs and waist. Hanging branches formed cool leafy bowers under which she stopped to rest. Eventually she continued in a tiny boat down a narrow waterway that passed through woods so thick no sunlight pierced the canopy. This little river ultimately spilled into an extraordinary body of water: a vast lake roughly the color and temperature of warm onion soup. It was like nothing Ida had ever seen before. The bare trunks of dead trees shot out of the surface as straight as totems planted by human hands. If not for the occasional barking dog or crowing rooster, she would have thought the place entirely uninhabited.

Ida had stumbled upon Sentarum, the vast floodplain of seasonal lakes and swamp forests that Tyson Roberts would go on to explore in 1976, and where my own attempt to find the Super Red was cut short on the tiny island of Pulau Melayu in 2009. To be fair, Ida didn't technically "discover" Sentarum, and not just because people had already been living there for thousands of years. Inspector L. C. Hartmann, working for the Dutch government, had come across the region in 1823, nearly three decades earlier. But when two more Dutchmen tried to return to the peculiar dark lakes, they found they had mysteriously disappeared. It was dry season.

Ida, however, almost certainly knew none of this history, and she was utterly transfixed by what she had found. Feeling drawn to the labyrinthine swamp, she wanted to stop and stay the night with the natives. But her Malay guide, that "rogue of a servant," refused—he was too terrified of the headhunters who lived there.

A HUNDRED AND sixty years later, in December 2011, I stood where Ida once did at the start of her expedition, surveying the wide, muddy mouth of the River of Man-Eating Crocodiles. "It's a good thing they only eat men!" my brother had e-mailed me. But the name is no joke. The saltwater crocodile, *Crocodylus porosus*, which inhabits the river, is not only the largest of all living reptiles but also the biggest terrestrial predator to roam the earth. At the Friendship Café in the sleepy fishing town of Lingga, I'd admired photographs of the infamous white-backed "Happy Bachelor," the Moby-Dick of crocodiles, who for ten years terrorized villages, devouring locals, while evading sharpshooters, shamans, and Crocodile Dundee types hurling grenades. In 1993, Dayak hunters finally blasted the giant beast out of the water with a volley of rifle fire. At nearly twenty feet long, and weighing over a ton, the dead carcass looked like a fallen tree with a crumpled smile.

No crocodiles were in sight, however, as I watched an isolated shower pour down like a bucket emptying on the opposite bank. It was rainy season, and I had returned to Borneo to reach Sentarum at high water as the arowana gathered to spawn. Would I find wild Super Reds in the tangled roots of the swamp? Or were they gone?

As I waited for the rains to refill the seasonal lakes in the interior, I'd traveled to the Malaysian state of Sarawak in northwest Borneo to attend a conference on arowana farming in the capital city of Kuching. When a friendly fisheries officer offered to let me tag along with two of his men going up the River of

Man-Eating Crocodiles, I jumped at the chance to trace the path that Ida Pfeiffer had pioneered in the 1850s—though I knew it was no longer possible for a traveler like me to reach Sentarum via this route.

I was actually closer to the swamp by more than a hundred miles here in the north than if I were approaching from the southwest as I had before. Yet two obstacles separated me from my goal: the same mountain range Ida had crossed barefoot, and something even more problematic—a stretch of the border between Malaysia and Indonesia that had been closed since an armed territorial conflict erupted between the two nations in the 1960s. A new immigration checkpoint was supposed to open soon, but word was that it hadn't yet.

WESTERN BORNEO

MALAYSIA

SARAWAK

Batang Lupar
(River of Man-Eating
Crocodiles)

Kuching•

South China Sea

Lingga•

Lubok Antu

Danau
Sentarum

INDONESIA

•Semitau

Kapuas

Pontianak•

KALIMANTAN

0 Miles   100
0 Kilometers   100

© 2016 Jeffrey L. Ward

In truth, I'd been drawn to Sarawak less for practical reasons than because of the romance of its history, which I'd come to appreciate as I dug deeper into the story of the arowana. Borneo's northwest coast was its busiest hub in the 1850s—the decade that saw the invention of the aquarium in England just as some of the most profound secrets of the natural world were being unlocked in the East Indies. Ida Pfeiffer wasn't the only adventurer to travel through this way. Someone else did too, the greatest field naturalist of the nineteenth century, and the prototype of the modern-day explorers I'd met.

IN 1852, THE same year that Ida reached Sentarum, a lanky, bespectacled young man stood bailing water out of a leaking lifeboat in the middle of the Atlantic Ocean. The ship carrying twenty-nine-year-old Alfred Russel Wallace back to London from the Brazilian Amazon had caught fire, and he watched in horror as four years of his work literally went up in smoke. Flames licked at the sails and swept through the cargo hold, where Wallace had stored nearly all the notes and thousands of specimens (several hundred new to science) that he had risked his life to collect. The only animal to survive was a single parrot.

Wallace spent ten days at sea in the open dinghy before he and the ship's crew were rescued by a passing vessel. By the time he reached London, he swore he was through with collecting. But it wasn't long before the itch returned. For his next destination, he chose the East Indies—the Malay Archipelago—because the islands were known to harbor rare and coveted animals. Unlike most gentlemen naturalists of his day, Wallace was of modest means, and he expected to support himself by selling specimens to museums and private collectors through an agent in London.

In the spring of 1854, Wallace landed in Singapore, where

he crossed paths with the White Rajah, James Brooke, whom Ida Pfeiffer had missed. Under the promise of Brooke's protection, Wallace decided to head to Borneo, arriving in Sarawak that November with the first rains of the season. He established a camp outside Kuching and set about collecting insects and animals—especially the "great man-like ape of Borneo," the orangutan. After shooting and stuffing a lactating mother, he hand-reared her infant, growing quite attached to his "little pet" and recording its remarkably humanlike tendencies.

During this time, nearly five years before Darwin published *The Origin of Species*, Wallace conceived of "the Sarawak Law," writing that "every species has come into existence coincident both in time and space with a pre-existing closely allied species." He was onto the idea of evolution. But he couldn't figure out the driving mechanism behind it—how the phenomenon actually worked.

Three years later, he was still in the East Indies, based now on the tiny island of Ternate, when he suffered a bout of malaria. Between paroxysms of fever and chills, his thoughts drifted to the same vexing questions: How did new species originate? What *was* a species anyway? Recalling a work he'd read a dozen years earlier—Thomas Malthus's *Essay on the Principle of Population*—Wallace considered how pressures such as disease and famine continually check the growth of humanity. It occurred to him that the same forces must act on animals, often in faster cycles. In his own experience, he had observed that within each species minute variations frequently afforded one individual a slight advantage over another—such as falcons with more powerful talons for gripping prey, or giraffes with longer necks for reaching higher foliage.

Suddenly it hit him in a flash: Those subtle differences, subjected to the pressures Malthus described, might produce increasingly specialized life-forms. The animals best suited to their environment survived while the less fit died—and thus new species came to exist. When Wallace's fever subsided, he quickly penned an essay entitled "On the Tendency of Varieties to Depart Indefinitely from the Original Type," then mailed it off to his idol Charles Darwin for his "perusal."

Did the wealthy aristocrat, a decade and a half older and vastly better connected than Wallace, steal his idea? Despite theories that Darwin may have borrowed some details, the answer is basically no. In a remarkable coincidence, it seems each man independently arrived at the same theory of evolution by natural selection. The idea that Wallace dashed off in a feverish fit, Darwin had been holding close to his chest for twenty years, fretting about the religious ruckus it stood to provoke.

Naturally, the older scientist was grievously pained at the prospect of having his life's work scooped. "All my originality, whatever it may amount to, will be smashed," he wrote his friend, the famed geologist Charles Lyell, who scrambled to arrange an urgent meeting at the Linnean Society. There, on July 1, 1858, Darwin's theory was presented first, with Wallace's essay tacked on at the end.

When Wallace, traveling amid islands half a world away, received a letter explaining what had happened, he could have been bitter. Instead, he was honored to have been included at all. He would spend another four years in the East Indies, eight in total, in part because he couldn't afford to go home. Having solved one mystery, he was about to open the door to another—this one starring the Asian arowana.

GRADUALLY, WALLACE BEGAN to see the distribution of animals as a puzzle that could be assembled to reveal the geological history of the earth. On the neighboring islands of Bali and Lombok, merely twenty-two miles apart, he was astonished to discover that the native animals differ far more than those of England and Japan. Eventually, he concluded that two distinct faunas inhabit the Malay Archipelago: one in the west, dominated by tigers, rhinos, and primates, and one in the east, characterized by kangaroos, koalas, and the spectacular birds of paradise. In 1859, he proposed a boundary running northeast through the narrow strait between Bali and Lombok—a division now known as Wallace's Line. He suspected the islands west of the line, such as Bali and Borneo, were once joined to Asia, while islands to the east, including Lombok, Sulawesi, and New Guinea, connected to Australia.

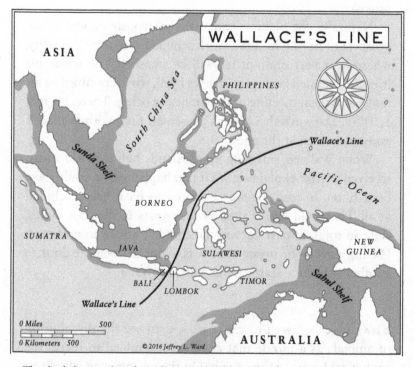

The shaded areas show how the major landmasses were connected during past ice ages when the sea level dropped as much as 650 feet.

Sonar mapping of the seafloor would eventually reveal that Wallace was right. His line traced the edge of the Sunda Shelf, a submerged landmass extending from the Asian mainland. During past ice ages, when sea levels were lower, dry land linked both Asia and Australia to the islands on their peripheries. But for the past 50 million years, the deep water between the two continental shelves had acted as a barrier dividing their distinct flora and fauna.

Wallace had forged a new field: biogeography, the study of the distribution of species through space and time. Eventually, he divvied up the entire globe into eight faunal regions that scientists still use today. Yet one nagging mystery surrounding Wallace's Line remained and would bedevil scientists for more than a century. Some of the strongest evidence for the divide came from the distribution of fish. For example, the freshwater species found in Borneo and those a thousand miles away in Australia did not overlap *at all*—with a single exception. What were arowanas doing on both sides of the line?

That arowanas inhabit Australia made sense because the fish was believed to have evolved on the supercontinent Gondwana, the bottom half of Pangaea, which once united Australia with all the other southern landmasses: South America, Africa, Antarctica, and Madagascar. When Gondwana crumbled, the ancestors of various arowana species drifted on continents across the southern hemisphere. But why was the arowana in Southeast Asia, which was never part of Gondwana? For a long time, the prevailing belief was that somehow the fish must have *swum* from Australia. Yet it didn't stand to reason that a species so highly adapted to freshwater could have crossed the salty seas.

In 2000, a pair of Japanese researchers finally solved the conundrum. By using the "molecular clock," the idea that DNA evolves at a relatively constant rate, they showed that the Asian and the Australian arowanas had last shared a common ancestor some 140 million years ago in the Early Cretaceous—far earlier than anyone had suspected based on how similar the fish looked.* But the real jaw-dropping revelation came when the researchers realized what exactly was happening to the surface of the earth at the time the species split: that's when the Indian subcontinent broke off from Africa, beginning its 100-million-year drift toward Asia, into which it ultimately crashed, creating the Himalayas.

The Asian arowana, it seemed, had caught a very slow ride across what is now the Arabian Sea, and then descended through the prehistoric waterways of Southeast Asia to modern-day Borneo inside the Sunda landmass. The fish was witness to the history of the earth—the shaping of the land beneath our feet.

BY CONTRAST, MY own attempt at an epic journey was proving rather tame. My first night on the River of Man-Eating Crocodiles, I curled up on the floor of a wooden shack alongside two barrel-chested fisheries officers snoring in rounds. Chong Ted Kin and Amri Bin Sulai had been crisscrossing the Sarawak countryside together for ten years to check on fishing licenses. Upon hitting their mats, they conked out immediately, and I tiptoed over them to turn off the light. Then I lay in the humid darkness beneath a lizard that cackled loudly from the ceiling—something I didn't know that lizards could do.

Other than difficulty sleeping, my experience going up the river wasn't proving to be much like Ida Pfeiffer's. Granted, a

---

* Arowanas aren't nearly as old as primitive fish such as lampreys and sharks, but they do represent something special. About 215 million years ago in the Late Triassic, a group of fish called the teleosts appeared and eventually became the most successful vertebrates on earth—today representing some 96 percent of all living fish. The earliest of these champions to roam the seas were likely the good old bonytongues, the group to which the arowanas belong. With their lithe bodies and prickly, daggerlike tongues, arowanas clearly found a design that works and have changed remarkably little ever since.

century and a half had intervened, but I was still surprised by how developed this part of Borneo was compared to the others I'd seen. For one thing, my assumption that we'd be traveling by boat turned out to be naive. The fisheries department didn't even *have* a boat, one of its officers had explained, because, "The water is very crocodile." Instead a road would allow us to cover the same territory in a day that Ida had traversed in nearly three weeks.

As we drove south the next morning, Chong and Amri began singing *gelombang, gelombang*, the Malay word for "wave," pretending for my benefit that we were being bounced around by choppy water instead of the deep ruts in the mud road. Out the window I watched rice paddies give way to the familiar vista of endless oil palms—the greatest threat to Borneo's biodiversity, alongside logging. "It's a golden crop now," Chong said, arguing that such development was good for the region. "Once you open up the forest, the wildlife may go further, disappear, or go extinct," he acknowledged. "These are the problems the environmentalists bring up. But I think that is not that important. We still have to earn money."

I thought about the old-growth forests of eastern North America, nearly all of which were chopped down at some point following the colonization of the New World. Something similar was happening in Borneo, though at a more rapid pace. At least 25 percent of the jungle had been clear-cut since the mid-1980s. Less than 50 percent of the "lofty virgin forest" Alfred Russel Wallace once admired now remains.

Not that Chong's backyard and mine were created equally. Tropical rain forests such as those in Borneo occupy only 6 percent of the earth's land area, but contain more than half its known species of organisms. The flora and fauna of the Malay Archipelago, in particular, are extraordinarily diverse. Part of the reason for this is that islands, because of their isolation, are hotbeds of speciation. It's no coincidence that Wallace and

Darwin reached their respective epiphanies while studying the animals of the Malay Archipelago and the Galápagos.

There is a parallel in the world of fish: Freshwater comprises only 2.5 percent of all water on the planet, most of it frozen as ice or buried deep in the ground. What remains, a mere one-hundredth of 1 percent of the earth's water, makes up all the rivers, streams, lakes, and wetlands in existence. Yet that tiny fraction supports nearly *half* of all fish species. Lakes are the aquatic equivalent of islands—natural laboratories of radical evolutionary experimentation.

Unfortunately, island species—and freshwater fish—are also more vulnerable to extinction, since they occupy smaller ranges and have evolved among fewer predators, competitors, and diseases. In 1967, the field that Wallace invented, biogeography, entered a new phase when the ecologists Robert MacArthur and E. O. Wilson coined the term *island biogeography* to describe the study of islands and the life they support. According to MacArthur and Wilson's model, when half the ecosystem of an island is destroyed, approximately 10 percent of its species will disappear. By the time 90 percent of the habitat is gone, about half the species are too. A prime example is what happened in Singapore, where over the past two centuries more than 95 percent of the original forest was cut down. The model predicts that the island nation should lose around 70 percent of its species of plants and animals. In real life, the actual number may now be as high as 73 percent, according to one study.

The Asian arowana was never native to Singapore. In the 1980s, however, after the country joined CITES, the government began to confiscate the fish from local aquariums and release them in the island's reservoirs, where a mongrel population has since bred and thrived. "We are caught in a very interesting dilemma," the freshwater biologist Peter Ng told me at the National University of Singapore, arguing that while native species require conservation, invasive ones should be eradicated or controlled. "I'm an old-fashioned biologist. I believe in a very simple mantra: terminate all aliens with extreme prejudice."

"Are you seriously thinking about *killing* them?" I asked.

"We are very confused," Ng admitted, adding that few precedents exist for such an invasive endangered species. The closest case he had come across was that of a possum protected in its native Australia, which was being systematically shot in New Zealand, where it had become a pest. No one could decide how to think of the Asian arowana anymore—as a precious mythical object or a mass-produced commodity or a dangerous invasive. Only one thing was clear: it was no longer just a fish.

AFTER ABOUT THREE hours in the car, we reached the small town of Lubok Antu at the top of the River of Man-Eating Crocodiles, where Chong and Amri planned to check out a tilapia farm. Chong told me Lubok Antu meant Ghost Lagoon. "It's a cowboy town," he said, explaining that it marked the frontier where migrant workers spilled into Malaysia from Indonesia. It was also reputedly a smuggling conduit for illegal timber along with exotic pets such as baby orangutans and arowana. Since the fish didn't live on this side of the mountains, any that came through here had to be originating from Sentarum, just across the border.

A few days earlier, however, back in Kuching, I had visited a much-publicized rescue center for arowana seized from smugglers and found a bizarre scene: a fish preserve with *no fish*. Eighteen months after forestry officials broke ground, the state agency possessed zero arowana. Not a single one had been seized. Now, as I stood at the supposed trafficking hub, I wondered, Were none being smuggled, none being reported, or none left at all?

So close to its headwaters, the river looked nothing like what Ida had described. A massive hydroelectric dam built in the 1980s had flooded the jungle with a vast turquoise lake, which shone in eerie contrast to the red earthen shore. Dead trees poked like skeletal hands out of the water where the forest had once been. Flooding from the dam had displaced some three thousand Dayak from twenty-six longhouses, forcing them to dig up their burial grounds in advance of construction. Now the single building that stood on the remote edge of the man-made lake was a posh hotel—Batang Ai Longhouse Resort, operated by Hilton, advertising the indigenous experience.

On a map, Sentarum looked to be right around the bend, but a formidable mountain range still lay between me and the swamp across the border in Indonesia. As I surveyed the rising emerald peaks and tried to imagine Ida striding across them, I was eager to get moving. The world she had seen didn't exist anymore—at least not on the Malaysian side of the mountains.

We drove to the new international immigration checkpoint that I'd heard about, but it wasn't yet operational. The construction site was deserted except for a single worker in a hard hat. Rolling forward through the free zone to an imposing locked gate, we stepped out of the car and gazed into Indonesia. Anyone could simply have walked around the gate, but I'd been warned that sneaking across the border could get you shot. Instead I climbed up onto the concrete ledge and dangled my arms through the bars like an orangutan, wiggling my fingers. The habitat of the Super Red was tantalizingly close, just thirty miles ahead. But I wasn't going to reach it this way.

CHAPTER TEN

Ghost Fish

INDONESIAN BORNEO—KALIMANTAN

In the end, I flew south to Kalimantan—Indonesian Borneo—where my quest had run high and dry two years earlier. The second week of December, I landed back in the port city of Pontianak and was immediately warmed by the sight of the man waiting for me at the airport—Hery Cheng, Kenny the Fish's supplier who'd traveled with me to Sentarum in 2009. Despite the steamy tropical heat, "Let It Snow!" was playing on his car radio.

"Does Santa come to Borneo?" I asked as we drove through town.

"Maybe *I* am Santa," he said, winking. The last time I'd seen Hery he'd been separated, but now he and his wife were back together and taking their three children to China for Christmas. This meant he couldn't return to Sentarum with me. However, I'd grown to recognize that traveling with a fish trader like Hery could be a bit of a liability anyway. Not only did local mistrust of the Chinese risk barring us from the lakes, but if we actually found a Super Red, no way would Hery just snap a photograph and toss it back in the water. Over drinks at the local mall, he tried to convince me to bring back any arowana

I might find. "They will be better off in my hands!" he said, his eyes glinting mischievously. He argued that the fish would live in a "mansion" (his ponds) where they'd dine on extravagant meals of cockroaches and frogs.

Because I wouldn't be traveling with Hery, I had arranged to get to Sentarum with members of a Pontianak-based NGO called Riak Bumi, devoted to protecting the lake region and helping its people—two issues often at odds. Merely four days later, I found myself in a little green longboat heading past the tiny island of Pulau Melayu, where my quest had ended in 2009. With me were three men from Riak Bumi: an earnest young adventurer named Hermanto, who'd accompanied me from Pontianak, as well as two locals—Pa Itam, a bachelor of sixty wearing a gray NEW YORK sweatshirt, and Jim Sambi, a wiry man in his thirties who'd grown up eating the Super Red for dinner. This time I would be seeing Sentarum with people who called it home.

Seated at the helm of the boat, Pa Itam skillfully steered us across the lake and into the swamp forest, a tangle of shrubs submerged in black water. Soon gnarled trunks rose high above the surface, a snarl of vines obscuring the view. Every direction looked the same. In the mid-1980s, the Dutch ecologist Wim Giesen, who was the first to study the lake region in depth, wrote in a report for the World Wildlife Fund, "For the best part of the year, the lakes form an intricate maze of waterways and open waters, and navigation is a hazardous task. Maps are not of much use." If Pa Itam was following a path, I couldn't make it out; but he had been traversing these waterways since 1970.

What is today known as Danau Sentarum National Park— some eighty-three black-water lakes connected by swamp forests—comprises one of the most species-rich lake systems on earth. Elegant waterbirds dart overhead. Hiding in the flooded forest are two of Borneo's most famous primates, the orangutan and proboscis monkey, as well as three distinct crocodile species. Still, Sentarum can feel deceptively quiet and eerily deserted, its face an inscrutable black mirror.

Come summer the water level drops as much as forty feet, as small rivers flow into the great Kapuas, draining the seasonal lakes so that it's possible to drive across the chapped floor of the swamp on a motorbike. During rainy season, however— which was well under way when I arrived that December— these rivers reverse course, acting as an overflow valve for the Kapuas. In the winter, more than 90 percent of Sentarum is inundated, creating the impression of a large inland sea.

Despite its inhospitable feel, the swamp actually has a long history of habitation. Evidence of human activity dates back more than thirty thousand years, based on carbon levels in the peat indicating a higher incidence of fires than climate fluctuations can explain. Today two main groups live in Sentarum— the largest being the Malays, the descendants of Islamic traders who traveled upriver in the seventeenth century and the native Dayaks they converted. Then there are the Dayaks who *didn't* convert to Islam, most of whom belong to the Iban tribe. Since the 1970s, the majority of these indigenous peoples have adopted Christianity. In our little group, Pa Itam was Malay while Jim and Hermanto were both Iban.

In 1982, Sentarum was designated a wildlife reserve largely on the basis of its value as a crucial habitat for the arowana. When Wim Giesen arrived there four years later, "it was actually sort of a blank thing on the map," he told me when I phoned the now-veteran researcher at home in the Netherlands. "I thought I was going to camp, but there was no dry land." Instead he lived on a houseboat, slipping into the eightysomething-degree water to swim or tow the boat. "It's very strange to be sweating in water," he recalled.

Giesen had been surprised to find some three thousand people residing in the swamp, many of whom had lived there for generations. Since that time, as illegal logging roads encroached on the area, the population had more than tripled. During fishing season, additional families arrived in houseboats from towns downriver. With the influx of people, the fish population had fallen dramatically. By the mid-2000s catches were down around 40 percent compared to the late 1990s.

From time to time, we passed clusters of homes built on stilts above the water. Pa Itam suggested stopping for the night at his niece's house, and just before dusk we reached a tiny village where the tin roof of a green-and-white mosque glinted in the setting sun. So close to the equator, there's hardly any twilight, and day flips to night as with a light switch. Darkness was already settling as we hopscotched across floating rafts and teetered up slippery planks. Just a year earlier, there hadn't been any electricity here; but now a generator growled to life, illuminating naked fluorescent bulbs that cast an eerie glow over the boardwalk.

Inside a small wooden house with a pointed roof and scalloped trim, Pa Itam plopped down on the bare floor in front of a television, hung his hat on his knee, lit a cigarette, and caught his grandnephew Adi Muhammad Akbar in a scissor hold as he passed. A news program was reporting that police had shot and killed four demonstrators protesting oil-palm plantations in Sumatra.

For dinner, Pa Itam's apron-clad niece brought out dishes of dried fish, stewed fish, fish crackers as crisp and chewy as pork rinds, and a delicate wild green plant called fish fern, which we ate seated in a circle on the floor. Before long, the room was crammed with what looked to be the entire village, curious to see outsiders. A wizened grandmother waved incense through the air to protect the baby in her arms.

"From mosquitoes?" I asked.

"No," Jim said, "from ghosts." Then he shrugged. "From mosquitoes too."

According to Pa Itam, this village was the first place outsiders came looking for the Super Red. When Chinese traders arrived in 1982, he himself had taken them here to survey the population. After that expedition, one of these men returned in a longboat hauling a generator and a film projector, offering to treat everybody to a movie that night. The entire village showed up for the big event. Meanwhile, the man snuck out onto the lake in his boat, collecting as many arowana as he could. Three decades later, the wounds were still raw.

"What was the movie?" I asked.

"*Rambo*," everyone answered in unison.

A somber man with saucer eyes shook his head ruefully and said something in Malay. "There were many, many arowana here," Jim translated. "Not a thousand, a *million*." Just a few years later, however, the Super Red had already been overharvested. According to Wim Giesen, by the time of his 1986 trip, finding one was like "finding a lump of gold." When he visited local communities, villagers would point out a new house or new boat and say that the owner had bought it after catching an arowana. Giesen never saw the wild fish himself.

By 1990, an arowana could fetch 1.2 to 1.3 million rupiah, more than $500 US, according to the villagers. They guessed that today it would go for five times that amount—if it could be found. But the species had all but disappeared. They'd caught one in 2007, one in 2008, and one in 2009. Since then, for nearly three years, they hadn't seen the fish at all.

THE NEXT MORNING, as we waved good-bye to Pa Itam's niece and set off down a little black river, I peered into the gently flowing water and wondered, If no one had encountered the wild Super Red in so long, was it gone for good?

There was one last place to look for an answer—with the Iban of Meliau Longhouse, who control the most famous arowana breeding ground in Sentarum.

Just as the blistering heat of the noonday sun began to beat down on us, we spotted our destination around a bend in the river. About the length of two train cars, the longhouse loomed high on stilts, a chaotic assemblage of gray wood and corrugated metal. Lines of laundry festooned the exterior like flags.

After docking at a floating raft and climbing up a rickety ladder, we peeked into the interior—a spacious, cool veranda that connected some thirteen doors, each housing a different family. A woman with a silver bun spotted Jim and called out merrily in Iban, "You got old!" Then she sized me up and asked, "Is she here for fishing?"

When Jim explained we had come to find the wild arowana, we learned there was big news. Two months back, an American angler named Robert had traveled through this way to catch snakeheads for sport. Notoriously aggressive, snakeheads are long, cylindrical fish with big mouths and shiny teeth. Native to Asia and Africa, they're terribly invasive outside their natural range—capable of surviving on land for up to four days and known to wriggle their way between bodies of water a quarter mile apart. In 2002, the fish appeared in a pond in Crofton, Maryland, presumably having wandered over from a live Asian food market. The subsequent hysteria inspired not one, but three, B-movies: *Snakehead Terror*, *Frankenfish*, and *Swarm of the Snakehead*.

So Robert had ventured all this way for the thrill of hooking such an infamous creature. Instead, on his third day fishing, he landed a giant Super Red the size of a snowshoe, thrashing and glistening. He had no idea what it was, but his guide, a young man from the longhouse, had the foresight to snap a photograph before releasing it. I examined the picture of a large, white-stubbled angler wearing a hat with a neck flap and holding up a gleaming orange-red fish, its mouth forced agape, gills flaring. The arowana's one visible eye glowered at the camera.

I wish I could say I was instantly happy to hear that the Super Red had reappeared, that my heart swelled with joy to know it wasn't extirpated from the swamp after all. For countless years, as all manner of humanity was pursuing the species, this one giant monster had been evading the lot of them, growing bigger and bigger and redder and redder. But my first reaction was:

He'd beaten me.

Not only that—I'd lost to *an American on vacation.* As I tried to digest this unwelcome news, I looked closer at the photo and saw that the fish appeared to have a spoon in its mouth. "Robert use a spoon," Jim explained, "to catch the fish."

"Why would the fish try to eat a spoon?" I asked.

"All fishes try to eat a spoon."

"But it's not *edible*," I protested.

This is maybe a good time to confess that I have never been fishing in my life. It is apparently Fishing 101 that many species will be attracted to a lure that glints in the sun, such as a metal spoon attached to a hook. Arowana are apparently no exception. I had no idea what to do with this information.

I had always heard the best way to hunt the species was to use a flashlight to search for its eyes gleaming near the water's surface at night. Naturally, I wanted Robert's guide—a young man named Sodik with a snarling wildcat tattooed on his back—to take me to the lakes that evening. When I asked, however, he looked pained and said he'd love to, if only it weren't Christmas Eve.

I was confused. It was December 14.

But Christmas, I learned, comes early in Sentarum. Like most of Borneo's Dayaks, the Iban of Meliau had converted to Christianity, and the entire swamp was served by a single preacher, responsible for bringing the holiday to some twenty-four villages many hours apart by boat. Each year, starting in October, the preacher worked his way from longhouse to longhouse.

Tomorrow he was scheduled to arrive at Meliau, and family members were returning for the holiday, many from Malaysia, where they worked for timber and construction companies.

Another problem with convincing someone to venture out after dark was that nobody searched for the arowana that way anymore. The odds of finding the fish had become far too unlikely to make the hunt worthwhile. In fact, the only fisherman who knew the old way of doing things was now blind.

Eventually, however, a tall, austere man named Badong agreed to take me, so long as I didn't plan on staying out until I actually *located* an arowana. After all, he hadn't caught one in ten years. Jim and Hermanto said they'd come along too, and late that evening the four of us set off cross-legged in the longboat, donning blue ponchos and rattan hats as a light rain began to fall. Soon clouds hid the moon and complete blackness enveloped us. Badong flipped on a headlamp that illuminated a tunnel through hanging vines and called out a warning over the roar of the motor: sometimes snakes, attracted to the light, dropped down from the trees. If that happened, he said, it'd be best to jump out of the boat.

In that moment, as I recalled what I'd read about the Asiatic reticulated python (the longest snake in the world at more than twenty feet), as well as lightning strikes, crocodiles, and the well-documented case of an orangutan raping a woman, I began to have second thoughts about what I was doing back in Borneo. My doctor had warned me not to immerse myself in the water, where a snail-born parasite could cause permanent paralysis. How much was I willing to risk to go after a fish I didn't even think was *good-looking*? By this point, the arowana inspired in me the kind of open disdain and irrational resentment with which Ida Pfeiffer typically viewed whatever she was imperiling her life to see. I was pursuing the Super Red like a sheriff after an outlaw.

Absorbed in these thoughts, and unable to see more than a

few feet ahead of the boat, I didn't realize we'd reached a lake until Badong switched off the motor and quietly rowed us into a mangrove. In the faint circle of light from his headlamp, I recognized the fanlike vegetation from the book that the venerable Kamihata-san ("explorer number *two* in the world") had given me at Aquarama. It was cassava—the plant that Iban women wove into mats, and that the men used for traditional penis ornaments (or so I'd read). It was also the plant that arowana sought during breeding season, its buggy roots providing an excellent habitat in which to hide, hunt, and cavort.

Badong shone the light into the lake, looking for arowana eyeballs, but the rain was coming down heavily now, streaming off the eaves of our hats. Despite the downpour, tiny winged insects swarmed around the lamp making it even harder to see. Squinting into the opaque, roiling water, I remembered how Heiko had told me he'd gone night-diving here with no equipment to speak of and shuddered at the thought. I felt certain he would've known what to do in this situation. I, however, was at a loss.

After nearly an hour scouring the mangrove with flashlights, rain pooling at our feet, Jim and Hermanto looked increasingly soaked and miserable. I wasn't going to be the one to give up first. But when Badong asked if I was satisfied, relief washed over me. "It's quite hard because the water level is already high," he told me using Jim as an interpreter. "It's not like dry season, when it's much easier." This was difficult to hear given that I'd already trudged through Borneo once during dry season only to be advised to come back during rainy season, after which I'd scrutinized weather patterns and turned logistical somersaults to return exactly as the waters were at their highest. Perhaps there was no good time to find the Super Red anymore.

That night, after changing into dry clothes, and then freaking out an entire family by removing my contact lenses, I bedded down on a soft floral mattress in their front room. As I lay reviewing the events of the day, I thought of Ida Pfeiffer and wondered if any skulls hung above me in the rafters. Jim had told me the longhouses still kept them—that he'd grown up with a whole roomful. While part of me totally agreed with my guidebook that it was "time to let the whole Borneo head-hunter thing die a quiet death," I was curious to see a few, and Jim had agreed to ask around.

But his inquiry came to naught. One set of seven had just been lost in a fire. Another collection belonged to a woman whose husband was away and who was too frightened to show us the relics herself. It wasn't the human heads she feared but rather a very special skull—that of a *naga*, the Iban dragon. When I asked what it looked like, she considered this for a moment, estimating the size with her hands, and then said something to Jim, which he translated: "Like a dog skull." At the last household he tried, the old woman told us that the skulls she had inherited from her father, who had inherited them from his father, were kept in a sealed box that she had never opened. Not once. Even her grandfather had been too afraid to look inside.

Drifting off to sleep, I jerked awake briefly when I heard someone approach. But it was just the grandmother of the family, who had come to tuck in the corners of the pink mosquito net she'd hung over me.

WHEN IDA PFEIFFER departed Sentarum in 1852 and began the long journey down the great Kapuas River, a strange new feeling came over her. After spending the past decade finding fault with nearly every culture she'd encountered across the globe, she realized that she actually *liked* the Dayak headhunters of Borneo and was sorry to be leaving them behind. She had found them honest, good-natured, modest, loving toward their children, and respectful of old people. "I should be inclined to place them," she wrote, "above any of the races I have ever known."

As for the small fact of their cutting off and preserving the heads of their enemies, she could not help but think that Europeans were just as bad or worse. "Is not every page of our history filled with horrid deeds of treachery and murder?" she wrote, suggesting that Napoléon could have decorated Versailles with the skulls of millions.

Ida would have liked to stay longer among the Dayak. Had she done so, perhaps she would have come across the Super Red arowana. She would not, however, have "discovered" the species, because the German naturalist Salomon Müller had already found the fish, albeit the green variety, in the southern Barito River sixteen years earlier in 1836. Müller's expedition through the East Indies was catastrophic. All four of his scientific colleagues died, three to disease and one during an insurrection on the island of Java. Indeed, of the eighteen researchers that the Natural History Commission for the Netherlands Indies sent to the region over thirty years, only seven made it back to Europe alive. Müller was fortunately one of them, landing safely in the Netherlands, where he delivered at least six preserved specimens of the Asian arowana to fellow zoologist Hermann Schlegel at the Natural History Museum in Leiden. In 1840, Müller and Schlegel published a description of the fish under the name *Osteoglossum formosum*, using the same genus as the South American arowana and choosing the Latin species name *formosum*, meaning "beautiful." (The species was later reclassified as *Scleropages formosus* after the discovery of the two Australian arowanas, which it more closely resembles.) "The fish in question is very common," Müller and Schlegel reported. "Its meat is not very nice and rather dry." The publication was a landmark—the very first fish from Borneo to be described.

Ida may have missed the Super Red passing beneath her petticoats, but she did deliver a collection of fish to the star ichthyologist of the century, Dutch military doctor Pieter Bleeker, who was stationed in Batavia, now modern-day Jakarta. In gratitude, Bleeker named one of the new species after her: *Engraulis pfeifferi*, the shorthead hairfin anchovy. Alas, it later turned out that someone else had already described the fish, and the dubious honor of being immortalized by an anchovy was wrested from Ida. Even back then, it could be hard to discover something genuinely new.

DARWIN'S UNWITTING COMPETITOR Alfred Russel Wallace had no such problem. After his eight years in the Malay Archipelago, he finally returned to London in 1862 with a staggering 125,660 specimens, more than a thousand of which were new species. Of all the beautiful things Wallace had laid eyes on, perhaps none moved him more deeply than the jewel-like king bird-of-paradise, which he encountered on the remote Aru Islands. Of the mesmerizing creature, cinnabar red with a pure white breast, a band of deep exotic green, and spiral-tipped tail wires about five inches long, he later wrote:

> I knew how few Europeans had ever beheld the perfect little organism I now gazed upon, and how very imperfectly it was still known in Europe. . . . I thought of the long ages of the past, during which the successive generations of this little creature had run their course—year by year being born, and living and dying amid these dark and gloomy woods, with no intelligent eye to gaze upon their loveliness; to all appearance such a wanton waste of beauty. Such ideas excite a feeling of melancholy. It seems sad that on the one hand such exquisite creatures should live out their lives and exhibit their charms only in these wild inhospitable regions, doomed for ages yet to come to hopeless barbarism; while on the other hand, should civilized man ever reach these distant lands, and bring moral,

intellectual, and physical light into the recesses of these virgin forests, we may be sure that he will so disturb the nicely-balanced relations of organic and inorganic nature as to cause the disappearance, and finally the extinction, of these very beings whose wonderful structure and beauty he alone is fitted to appreciate and enjoy. This consideration must surely tell us that all living things were *not* made for man.

It was a prescient vision of the world to come, when the phrase "wanton waste of beauty" would no longer evoke the bird undiscovered so much as the many creatures doomed by their very loveliness. The subject of beauty was a point of contention between Wallace and Darwin, because Wallace never accepted Darwin's theory of sexual selection, the idea that competition for mates could result in such marvels as the peacock's tail. Instead, Wallace believed that coloration was determined by the laws of utility and primarily a matter of camouflage or warning. For Wallace, evolution never fully explained beauty—certainly not the level of beauty that he had seen. (Neither man anticipated E. O. Wilson's biophilia hypothesis: that beauty might evolve in the brain of the beholder, that the human love of other life-forms might be an instinctive bond between *Homo sapiens* and nature.)

Though Wallace's journey to the East Indies was his last, he spent the rest of his life elaborating on what he had learned during those years. Living to the age of ninety, he witnessed the era of the automobile and the airplane while becoming something of a horse and buggy himself. His kind—the self-taught scientist, the independent naturalist—was rapidly dying out.

MY OWN TIME in Borneo was nearing an end. The day before heading back to the coast, having said good-bye to Pa Itam and returned to dry land, I hired a driver with Jim and Hermanto to visit the old Iban longhouses dotting the hills east of Sentarum. As we bumped along the mud road in the morning sunshine, we passed an ancient-looking man hobbling along the shoulder like an apparition. He had a stringy gray beard and was wearing a trucker's cap set high atop his head. His bare legs, poking out of shorts, were as knobbly as the walking stick on which he leaned. Recognizing him as the headman from a nearby longhouse, the driver pulled over to offer a ride. As the old man climbed into the backseat, flashing a toothless grin, his face corrugated with wrinkles, I noticed that his throat was etched with the tattoo of a tree. The earthy smell of his sweat filled the car.

A few minutes later, we reached the longhouse where the headman lived. Built on dry land at the base of a hill, it seemed very old. Most of the newer longhouses were constructed from skinny gray boards that bowed under the weight of footsteps, but this one featured the kind of thick dark planks no longer possible to find. A battery-operated radio was playing a techno version of "We Are the World." On the veranda, a group of young men with tattoos and piercings—half-traditional, half-punk—sat shading a map of the park to indicate the intricate quilt of protected zones.

"How long has this community been here?" I asked Jim.

He relayed the question to the old headman, who replied at least 100 years, maybe 150, but he wasn't sure. He said he was born in 1914. At ninety-seven, he could still walk to and from the nearest town, a three-hour round-trip drive, in a single day. The trick, he maintained, was not eating fried foods. "If you eat a lot of cholesterol," Jim translated, "your knees will be fat."

"So did he used to eat the *siluk*?" I asked, using the local name for the arowana.

Recognizing the word, the headman nodded vigorously. "A *lot*," Jim translated. "An arowana meal is like a multivitamin!"

I wanted to know how life in the area had changed over

time. The headman thought about this and said that the tribes used to engage in a great deal of infighting. His longhouse had once been decorated with many skulls. During World War II, with the encouragement of the British, Dayak warriors beheaded Japanese soldiers occupying Borneo. Now the old heads were mostly kept at a cemetery three hours' walk into the jungle; only a single skull remained in the longhouse. These days the people were fighting a different war. "For the protected area," he said, indicating the young men and their patchwork map. The oil palms were encroaching upon their land.

The headman led us to the doorway of an old woman who beckoned Jim, Hermanto, and me to climb a steep notched log that ascended to the rafters, which were piled high with rattan baskets and rolled-up mats. Dust streamed through the beams of sunlight as we sat down to a plate of puffed rice and a bottle of rice wine called *tuak*, which the woman served in flower-painted tin cups. Meanwhile, her grown son disappeared up another ladder and returned holding a metal platter bearing a blackened skull encircled in a dried vine like a crown of thorns.

Waving her fingers over the head, the old woman began to murmur a prayer. Jim whispered that she was telling the spirit that the weather was nice out today so there was no reason to curse us. Hermanto's eyes grew large, and he disappeared down the ladder, peeking his head up through the hole to watch from a distance. Next the woman sprinkled the skull with puffed rice, which mostly fell in its eye sockets. She spilled liquor over the ghastly grin and took a sip herself. Then she pointed at me and indicated I should do the same.

As I poured *tuak* from my cup onto the chipped black teeth, I asked who this person had been. The old woman told Jim she thought the skull belonged to a Japanese soldier killed by her father. But she couldn't be sure. Her father had left behind five skulls, of which her sister had buried four. No one knew which was which anymore. They hadn't written any of that stuff down.

"Why did she keep this one?" I asked Jim.

"She said it's like a symbol," Jim said, "like a family treasure, something left over from another time."

I had been on a treasure hunt myself, pursuing a fish that also seemed to belong to another time. The Asian arowana was not extinct—it was in tanks all over the world. But its eons of life in the wild seemed almost at an end. Although the Super Red hadn't been entirely extirpated from Sentarum, it was so incredibly rare that when one turned up, it was the talk of the swamp. No one was eating it as a supervitamin anymore.

True exploration required untrod ground. My thoughts drifted some twelve hundred miles northwest to the obscure jungle rumored to hide that holy grail of fishes, the batik arowana in Myanmar. Had an unknown species really been hidden from humanity all these years? The booming voice of the grand old man of ichthyology, Tyson Roberts, rang in my head, cajoling me to go after the fish: "If you're a real traveler, you could do something zoological for me. And I'll publish it if I get it. I'll live long enough to publish."

BACK ON THE veranda, the old headman folded himself onto a mat and urged me to take a seat across from him. He locked his sparkling dark eyes with mine, chattering away in high, nasal syllables.

"They tell a story about the beginning of the earth," Jim translated during a pause. "We are all the same thing. The fish. The forest. The animals. All are the same. Arowana too."

Sure, it was a cliché. But the man was 100 percent right. Wallace and Darwin had taught us that the "natural relationships" Linnaeus described were actually literal relationships— branches in a giant family tree. Flapping around on dry land, we were simply another line of fish that had out-adapted our forerunners and come to dominate the biosphere like never before.

Then the headman launched into a second story, this one far more recent. It was about a stranger who had appeared in a boat at Meliau three years ago with a fishhook stuck in his cheek. After a traditional healer had been summoned to remove the hook, everyone suddenly realized the stranger was not a person at all but rather the king of fishes.

"Wait," I ventured, when the story concluded, not sure where to start. "This happened *three years ago?*"

Jim translated the question, and the headman nodded vigorously.

Then he leaned in close so his face was just inches from mine and began to speak with great urgency, eager to impart one final emphatic point. "Arowana is just like ghost—he can disappear," Jim translated. "The arowana is not extinct but *hiding*."

PART IV

FISH #32,107

CHAPTER ELEVEN

Monsters of Our Making

NEW YORK → TOKYO

The great American ichthyologist Tyson Roberts had disappeared. I hadn't heard from him in six months, since he sent me a viral video of a beachgoer strolling into the surf and getting snatched by a killer whale ("If it is fake it seems very well done"). That was in the spring, not long after our one and only phone conversation. At the time, Tyson had been convalescing in Panama, stranded half a world away from his home base in Thailand.

Now it was January, I had just returned from Borneo, and Tyson had stopped responding to e-mails. I tried calling his old number in Panama, but the line was disconnected. He'd been fretting about having to vacate his temporary housing and hadn't known where he'd go next. But I doubted he could've gotten far. Due to his deep-vein thrombosis, a long flight could kill him—the pressurization blasting a blood clot straight to his lungs.

It was urgent that I get hold of Tyson because I was preparing to leave for Myanmar in a month to collect the batik arowana for him, so he could declare it a new species (if it turned out to be one). Time was of the essence. Not only was rival ichthyologist Ralf Britz at the London Natural History Museum after the same fish, but the German aquarium magazine *Amazonas* had just run two alleged photographs of it. The pictures revealed an arowana imprinted with an intricate bronze pattern at once irregular and repeating. "Ever since we heard the first rumors of these fishes we have been trying to get more information," the text read in German. "The fish's precise provenance remains a secret."

For thirtysome years, Tyson had been a research associate—a kind of ichthyologist-at-large—at the renowned Smithsonian Tropical Research Institute in Panama City, the only bureau of the Smithsonian based outside the United States. Yet the Smithsonian didn't know where he was either. One administrator said she thought he had left for Suriname on the northeastern coast of South America. But there had been no word from him in months. He had left no phone number or forwarding address. His mail was piling up. Was the grand titan of ichthyology all right? Or was I fetching a fish for a dead man?

HEIKO, AT LEAST, I knew was alive. He had bounced back with superhuman speed from his horrific crash in the Heikomobile. However, with less than a month to go before our expedition to Myanmar, he still wouldn't commit to firm dates. All I could get out of him was that he expected to fly there from India sometime in February. I had no idea how he would pull off a last-minute visa, given that I'd had to apply for mine three months in advance with a round-trip plane ticket in hand, not to mention a nom de guerre. But he didn't seem worried. For my part, I'd been granted the maximum stay that the Myanmar government permitted, four weeks, spanning the month of February. So if Heiko materialized, I would be there to meet him.

Most nights, I lay awake worrying that Heiko would stand me up. Other nights, I worried about the exact opposite. Why had I thought it was a good idea to venture into a remote jungle pocked with land mines? For starters, I had no idea what sort of equipment to bring. PowerBars? Antivenom? I didn't even know if I should pack a tent. In Borneo, I had stayed with the locals, sleeping amid bags of rice in supply closets or lined up on the floor with everyone else. But what if there weren't any locals? I had a vague memory that Heiko slept out in the open without a mosquito net. I was not really that type of person.

Even if I were, curling up on the jungle floor didn't seem smart in Myanmar, which has one of the highest incidences of lethal snakebites in the world. Virtually all the country's venomous snakes are most active at night, according to "The Dangerously Venomous Snakes of Myanmar," which includes an appendix on the treatment of snakebite by Joe Slowinski, curator of herpetology at the California Academy of Sciences, who subsequently *died of a snakebite in Myanmar*. In 2001, Slowinski was surveying reptiles and amphibians in the foothills of the Himalayas when a many-banded krait bit his finger, injecting him with a powerful neurotoxin. Over the next twenty-nine hours, as fourteen fellow scientists and four Burmese field assistants performed artificial respiration and radioed for a rescue helicopter that the government failed to send, Slowinski gradually succumbed to paralysis until he spelled out his final message: "Let me di."

King cobras. Spectacled cobras. Monocled cobras. Myanmar has so many snakes that some don't even have names yet. Rendering such hazards even more dangerous is the general lack of infrastructure and telecommunications. I read that the country had fewer cell phones per capita than North Korea. Even if I could get my hands on one, my conversations would almost certainly be monitored by military intelligence, and the phone wouldn't work in the remote south where the fish was supposed to be. Jeff wanted me to take a satellite phone. But satellite phones are verboten in Myanmar, and breaking rules makes my feet sweat from anxiety. More than anything else, I worried that Heiko would (as others had warned me) take some crazy risk and land us behind bars. Like his mother, after all, he had ended up on the wrong side of the law in the past.

IRONICALLY, THE ONCE call-evading Ralf Britz at the London Natural History Museum was the only one I could reach on the phone. He seemed to be warming up to me, but he still wouldn't put me in touch with his contact in Myanmar, an expert in the native ichthyofauna. "I just don't know you," Ralf said warily. "If we could meet up for a beer . . ."

Actually, I had to admire Ralf for proceeding with caution. The more we talked, the more I developed conflicted loyalties. I felt guilty about potentially stealing the batik arowana out from under him. I understood his hunger to find the fish—I felt it too. I'd been reading so much Alfred Russel Wallace that my head had gotten stuck in the Victorian era and caught up in the romance of exploration. Unlike the Super Red, from which I'd now moved on, this unknown species represented a chance to make a true discovery—a rarity in our modern world. It seemed almost providential that the opportunity had fallen into my lap: privately, I indulged the fantasy that this prehistoric creature, its scales scrawled with cryptic calligraphy, had been waiting all these years for me to find it.

If Heiko came through and met me in Myanmar, any arowana he caught would naturally go to his old friend Tyson, who had first asked me to retrieve the fish. But if Heiko stood me up, and Tyson never resurfaced, then what? I needed a backup plan, and I broached the topic delicately with Ralf. "Would you potentially be interested in getting the fish?" I asked after an uneasy pause, instantly feeling like the two-timer I was.

"Of course, but I don't see how that's going to happen, really." Unlike Tyson, Ralf had zero faith in my mission—he seemed to view it as almost comically doomed. "I think the chances of you going down to the Tenasserim are small," he said bluntly, pointing out that getting a permit from the military to head south took months. From there, I'd have to ascend into the hills. "And really, I would be very surprised if you got permission to go up into the mountains. Very, very surprised."

When I told him I figured I'd start by introducing myself to the Department of Fisheries, he issued a sharp gust of laughter. "You're really thinking too much in American standards," he said, describing a formidable gate and menacing guards who didn't speak English. In Ralf's opinion, I wouldn't get anywhere without a knowledgeable contact on the inside. "Maybe you should book your flight via London," he said, presumably so we could have that beer and he could size me up in person.

But my itinerary was already set, and I couldn't change it without messing up my hard-won Myanmar visa. Besides, I had picked my point of transfer—Tokyo—for a particular reason. Now I asked Ralf if he had any interest in koi.

"No," he said flatly, with characteristic cynicism, "because they're not really fish."

NOT REALLY FISH? What is a "fish" anyway? Surprisingly, the word has no clear scientific meaning. Birds, mammals, reptiles, amphibians, and insects all comprise classes within the kingdom Animalia. But the group we call fish lumps together a hodgepodge of radically different organisms belonging to a variety of classes. Every so often an ichthyologist proposes a new way to define what a fish is—usually some variation of a water-dwelling, cold-blooded vertebrate with gills, scales, and fins. Yet exceptions abound: most swamp eels lack scales and fins; tunas and certain sharks are warm-blooded; and lungfishes are far from the only fish with lungs or other means of breathing air in a pinch. (Even arowanas do this.)

I'd encountered the question of what qualifies as a fish in a different context when I learned that captive arowana don't count in an ecological sense. (Recall the mantra of conservation biologists: "A captive fish is the same as a dead fish.") Ralf was making a somewhat related point. He was talking about artificial selection, that age-old form of genetic engineering, which Darwin studied to understand how animals could be shaped through time. Could selective breeding transform a fish so radically that it became something else? Dogs, after all, are no longer wolves. House cats aren't wildcats. Pigs aren't boars.

When Ralf said that koi were "not really fish," he was pointing out their artificiality. By his definition, true fish were necessarily wild, the beautiful and successful products of evolution by natural selection, elegant embodiments of form following function. The fact that I was so keen to find the wild arowana, that I thought tracking one down would somehow be different from seeing someone's pet, suggested I might agree with him. I knew Heiko did. I remembered his disdain for the garish varieties of selectively bred discus at Aquarama, which he dismissed as "terrible monsters."

Darwin himself wrote that the eighty-nine varieties of goldfish he had counted "ought to be called monstrosities." They were the only fish he knew of to have been radically modified by artificial selection. With their grotesque bauble heads, bubble cheeks, and telescope eyes, goldfish proved, he wrote, that "it is an almost universal law that animals, when removed from their natural conditions of life, vary, and that races can be formed when selection is applied."

Unbeknownst to Darwin, another such man-made creature was emerging in his day: the Japanese koi. At Aquarama, Heiko had told me that koi—not arowana—rank as the world's most expensive pet fish. (Arofanatics get around this by calling the

Asian arowana the world's most expensive *aquarium* fish, since koi are kept in ponds.) Heiko claimed he knew of a koi that sold for $3 million. When I asked what was so special about it, he grew impatient and said it was just part of a different tradition: Koi were auctioned off like works of art, each one photographed from every angle and featured in a glossy catalog. The owner of the winning fish would become famous throughout Japan. "Every single person will know him," Heiko said. "When someone gets that koi, he can breed it like a racehorse. They have a pedigree. You don't have this—yet, at least—with the arowanas."

But that was where the dragon fish seemed to be headed. According to Ralf, its transformation had already begun—and would speed up dramatically if Kenny the Fish's new laboratory succeeded in producing "tailor-made arowana." Hence I planned to stop in Tokyo to check out Japan's largest koi show and investigate what happens when a wild fish is domesticated—when it transforms, so to speak, from a wolf to a dog.

Right before departing New York, I finally heard from Tyson, who was, despite my fears, very much alive. His note was short and somewhat mysterious: "Yes, I am in Suriname. Good to hear from you. I have to leave immediately, no time for a more detailed reply. I shall get back to you right away about Tenasserim River."

A few days later, a longer but no less cryptic message appeared from which I slowly gathered that Tyson had *no idea who I was*. All this time, I had been preparing to do his bidding in the name of science. Meanwhile, he'd forgotten that we ever spoke—much less that he'd told me about the rumored arowana in Myanmar.

Looking back through our correspondence, I came across the viral video he'd sent me of a man strolling into the surf and getting snatched by a killer whale. I did a little poking around online and realized it was most definitely a fake, an ad for a department store in the Dominican Republic. If Tyson had been duped once, maybe the batik arowana was just another hoax. Suddenly I felt that I was embarking on a fool's errand of colossal proportions, but it seemed too late to back out.

LONG BEFORE I was ready, I found myself hugging Jeff good-bye at the airport. Thirty-six hours later, I had splashed cold water on my face and stood shivering in a vast exhibition hall in Tokyo's Ryutsu Center, watching a koi named Mika drift across a blue canvas vat the size of a kiddie pool. Mika was four years old, two and a half feet long, and white with cherry-red splats across her back, as if someone had tripped and spilled ketchup on her.

She certainly looked and acted the part of a fish: scales, fins, gills. Her primary activity was to drift across the little pool until she bumped her snout on its canvas edge, at which point she flinched in surprise—*Whoa, a wall!*—altered course, then drifted until she bumped her snout again. Was this normal koi behavior? Or was Mika the victim of all the inbreeding in her lineage? Among carp, she was royalty, the daughter of New Smile, whose oversize portrait hung on the wall amid the forty-three past grand champions of the illustrious All Japan Koi Show, held every year since 1968.

Unlike her mother, Mika hadn't snagged any major prize. "What hurt her this year is she lost a scale," said her owner, Rick Costantino, a chemical engineer from Seattle, pointing to a little pink spot defacing one of her red splotches. Like the other competitors, Mika had spent the year boarding at a koi hotel in the Japan Alps, and Costantino had only recently gotten reacquainted with his fish. He was disappointed to find she'd had a big growth spurt, propelling her into the eighty-centimeter division, like a wrestler at the bottom of a weight class. "She's competing against fish that are *two inches longer*," he said significantly.

More than fourteen hundred koi ranging in size from a gerbil to a bulldog drifted across identical blue pools dotting the exhibition hall. Each fish was painted in a palette of red, white, black, gold, and/or blue. Some had scales that glistened like diamonds or were outlined like mesh. Others were "naked," with no scales at all. The grand champion looked a bit like a Jackson Pollock in red, white, and black. According to *The Cult of the Koi* by Michugo Tamadachi, it could easily sell "for the value of a Rolls-Royce motor car." But no one I spoke with had heard of a winner fetching more than a few hundred thousand dollars, in the supposed ballpark of the most expensive arowanas.

To me, the koi were simply cute, with their sweet cheeks and poochy tummies. If anything about them was extraordinary, it was that all the hundred-plus brilliantly colored varieties arose within a century from the snot-colored common carp, one of the dowdiest, least charming fish ever to grace the waterways of earth.

MERELY TWENTY THOUSAND years ago, all animals were wild, with the possible exception of a few tamed wolves gradually becoming dogs. Then, about twelve thousand years ago, came the Neolithic revolution, which saw the widespread domestication of livestock such as sheep, goats, and cattle. Today, according to one calculation, farm animals make up about 65 percent of all terrestrial vertebrates by weight, with humans and our pets accounting for most of the rest. Wildlife comprises only 3 percent of the total. As the zoologist James Serpell of the University of Pennsylvania writes: "Within the space of a few thousand years, domesticated animals have virtually replaced their wild progenitors across most of the planet's surface." Water, however, has long been a different story—the last true domain of wild things.

The first people to keep fish in artificial pools were probably the ancient Sumerians who settled in Mesopotamia (modern-day Iraq) more than six thousand years ago. Fish stocked the first aquifers, which fed the first irrigation system, which produced the first grain surplus and a corresponding population boom that led to the birth of the world's first city and first written language. The ancient Egyptians, too, reared species such as Nile tilapia in ponds. But we can't know for sure whether any of these fish were truly domesticated—that is, whether they evolved in response to selection pressure associated with a life under human supervision. Or if they were just "exploited captives," meaning (in the lingo of ethologists) that their biological identity remained intact, unchanged from their wild counterparts'.

The large-scale farming of fish, known as aquaculture, was probably invented in China over three thousand years ago, with the rearing of various carp species in flooded rice paddies and in ponds on silk farms. Around the dawn of the first millennium, the Romans began keeping the common carp from the Danube River as a culinary delicacy. Then in the twelfth century, farming the species became big business throughout Europe. Gradually, the fish began to change as it naturally adapted to its man-made environment: Its torpedo-shaped body compressed, growing plumper and stouter. Its mouth widened, its intestines lengthened—all evolutionary changes that enabled it to thrive in crowded ponds.

Eventually, this domesticated carp spread eastward all the way to the Japan Alps, where farmers stocked the species as a source of protein through long, harsh winters. Buried beneath twenty feet of snow for much of the year, the fish lived in total darkness, possibly triggering the appearance of the colorful mutants first recorded in the early 1800s. Over the years, the farmers separated these red, yellow, and white fish from their homely brethren and selectively bred them, eventually producing *nishikigoi*, meaning "embroidered carp," or *koi* for short.

At the turn of the twentieth century, koi and goldfish were

among the only domesticated fish in the world. Then, in the early 1900s, Siamese fighting fish and guppies subjected to intensive artificial selection were soon available in a wide array of shapes and colors. By 1935, the most popular aquarium guide listed eight domesticated species. By 1955, there were ten. Today more than fifty species of fish have been selectively bred as pets and are available in well over three hundred varieties.

This proliferation of domesticated breeds has taken place within a much larger context: the explosive growth of aquaculture since the 1950s, which has seen a more than seventy-fold increase in the production of farmed fish for consumption. Thanks to this "Blue Revolution," people are able to eat a greater amount of fish than ever before, even as some 90 percent of the world's marine stocks are at maximum production capacity or overfished. In 2009, when I first visited arowana farms in Southeast Asia, farmed fish stood to surpass wild catches for the first time in human history, according to the United Nations. Many of the 250 species reared in ponds now lie somewhere on the blurry continuum of domestication.

As I stood watching the koi putter across their pools in Tokyo, I was beginning to understand why the arowana had come to represent the wild for me. Fish have long been our last wild food and, together with reptiles and amphibians, our last wild pets. In recent decades, however, that has been changing. I wondered how long it would be before the aggressive arowana grew plump and complacent like a koi.

Domestication disrupts the normal behavior of fish. They become chubbier and weaker with smaller brains. They lose their natural instincts and competitive edge. Scientists have found, for example, that when domesticated salmon escape into the Atlantic and mate with wild salmon, their offspring are unable to find their way back to freshwater to spawn. Eventually, experts warn, such interbreeding could lead to the extinction of wild populations.

On Kenny the Fish's farm, Alex Chang had argued that preserving the Asian arowana in captivity leaves us the option to reintroduce it into nature someday. But it doesn't work like that. Take what happened to the common carp: Today the IUCN ranks the fish among "the world's 100 worst invasive species," while simultaneously listing it as vulnerable to extinction. This seeming paradox arises, in part, because the carp found in nature are now often feral domesticates. Meanwhile, the original wild-type fish is disappearing—going the way of the majestic aurochs (from the German *ur-ox* or "proto-ox"), the progenitor of the modern cow, driven to extinction in the seventeenth century.

ON MY LAST night in Tokyo, after I left the koi show and made my way back to the hotel through a swirl of futuristic lights, I thought about how the city was a fitting place to have found my post-fish, the koi. But I'd had my fill of man-made creatures in a man-made world. Now I was eager to travel to a country locked in the past to search for an "ur-fish," a proto-fish, one that did not yet have a name.

On my last morning in Japan, I awoke for my flight to Myanmar anxious about the adventure ahead. When I rolled over to check my e-mail, I found—at last—a message from Heiko: He was coming. He would see me in Yangon in two weeks.

CHAPTER TWELVE

"The Authorities Will Be Watching You"

MYANMAR

There was nothing beneath me but green. It spooled in gentle undulating waves under the white rays of the morning sun. From Tokyo, I had flown six hours to Bangkok. Now I was on a short flight into Myanmar, where we would soon be landing in the former capital, Yangon. I realized I must be looking down on the Tenasserim Hills, the low mountain ridge older than the Himalayas, which forms the geographical tail of the country. It was February 2012, and there were 32,106 fish species known to exist in the world. I pressed my nose against the window and wondered if number 32,107 was waiting for me in the vast jungle below.

Compared to its neighbors, Myanmar still has a relatively high amount of intact forest, largely because its turbulent history since World War II impeded development. Before that, the British governed the country—then called Burma—as a province of India. Despite exploitation and neglect as a colonial backwater, it was one of the wealthiest regions of Southeast

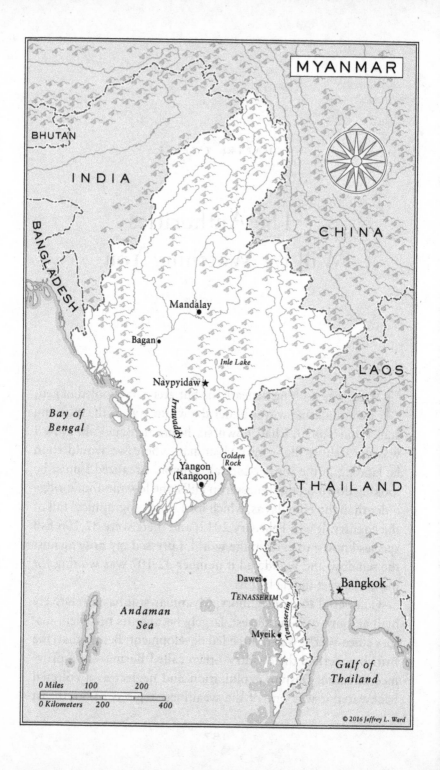

MYANMAR

BHUTAN

INDIA

BANGLADESH

CHINA

Mandalay

Bagan

Inle Lake

Naypyidaw ★

Irrawaddy

LAOS

*Bay of
Bengal*

Golden
Rock

Yangon
(Rangoon)

THAILAND

Dawei

TENASSERIM

Bangkok ★

Myeik

Tenasserim

*Andaman
Sea*

*Gulf of
Thailand*

0 Miles   100   200
0 Kilometers   200   400

© 2016 Jeffrey L. Ward

Asia. But the Japanese occupation left it a shambles. After the war, under the leadership of General Aung San, founder of Burma's Communist Party, the country achieved independence from the British in 1948—six months after the general himself was assassinated by a rival. Without its beloved leader, the new nation struggled to find its footing. In 1962, a military takeover set in motion a miserable Marxist experiment called the Burmese Way to Socialism. By the 1980s, the country was one of the poorest and most isolated in the world. The ruling general, Ne Win, a former post office clerk known for his superstitious beliefs in numerology, made matters worse when he spontaneously demonetized certain banknotes, replacing them with denominations divisible by nine. It was impossible to make change. People's life savings evaporated overnight.

All this led to tremendous dissent. In 1988, students across the country launched a massive demonstration against the oppressive regime. The government's response was brutal: soldiers opened fire with machine guns, killing more than three thousand. The next year, as if to wipe clean the slate of history, the military junta not only changed the name of the country from Burma to Myanmar and the name of the capital from Rangoon to Yangon, but also renamed various cities, towns, streets, rivers, and mountains.

Little improved in the years that followed. A different general, Than Shwe, seized power in 1992. Famously reclusive, he moved the capital from Yangon in 2005 to a malaria-ridden plot called Naypyidaw (Abode of Kings), some two hundred miles to the north. Meanwhile, Myanmar continued to have the lowest per capita income in Southeast Asia—lower than that of Laos or Cambodia.

After the government raised fuel prices in 2007 with calamitous consequences for the country's poor, thousands of monks took to the streets in the Saffron Revolution, so-called for their rust-colored robes. Though Myanmar is devoutly Buddhist, the military did not hold back. Monks were beaten unconscious and shot with rubber bullets, then real bullets. It took three days to suppress the protests.

When General Than Shwe finally decided to retire in November 2010, his handpicked successor, former military commander Thein Sein, won an election that the United Nations condemned as mere pageantry. Little more than a year later, when I stepped off the plane into the muggy heat that blurred the gilded facade of the airport, I expected to enter an Orwellian dystopia.

Instead, inside the air-conditioned arrivals hall, I found a massive crowd of foreigners backed up at customs: European backpackers, American diplomats, Chinese businessmen, and Singaporean bankers. I had arrived with the Burmese Spring.

WITH THE BENEFIT of hindsight, it was possible to see the first shoots of green peeking through the snow as early as the previous March, when the new president, Thein Sein, acknowledged problems in the country such as "the hell of untold miseries" along the war-scarred borderlands. Then in October (as I made plans with Heiko to go after the batik arowana) Myanmar's parliament legalized trade unions, banned since 1962. At the end of November (as I ventured to Borneo) Hillary Clinton became the first American secretary of state to visit the country in over half a century. A mad rush of investors followed, straining Yangon's infrastructure to the max. All hotels were booked solid. From the outside, experts debated whether change had truly come, and if it would stick. On the inside, tourists were arriving en masse.

Almost all these sightseers were traveling a well-prepared, government-approved route, a loop that took them up to the last royal capital, Mandalay, then down to the misty waters of Inle Lake, and finally west to the ancient city of Bagan, where thousands of temples glitter in the heat of the central plains. No one else, it seemed, planned to go south, at least not beyond the Golden Rock, a precariously perched granite boulder said to balance on a single hair of the Buddha.

Despite stretching a thousand miles, the Tenasserim region (also called Tanintharyi) merited only half a page in my guidebook as it was so difficult to access. In modern times, Tyson Roberts was the only ichthyologist to have collected fish from the Great Tenasserim River, sneaking in from Thailand with the help of rebel Karen boys armed with Kalashnikovs, who carefully picked their way through the jungle to avoid land mines. These reckless missions restricted to an upper stretch of the river were far from exhaustive.

Since Tyson had found no arowana, Heiko and I would need to reach the lower Tenasserim. I had no idea if that was possible—Ralf Britz certainly hadn't thought so—but I tried to banish doubt from my mind as I stood in the customs line at the Yangon International Airport. Even amid the mob of foreigners, I worried about being exposed as a journalist. If anyone asked, I wasn't sure how to explain why I'd packed audio recorders, notebooks, camping equipment, and maps of a war zone. To add to my nerves, I was carrying so much cash that it was too thick for my wallet and was stashed in various nooks and crannies upon my person. No credit cards or traveler's checks were accepted in the country, and ATM machines didn't exist. There was no way to get money once inside.

Exacerbating my paranoia was all that I'd read about Myanmar's notorious surveillance machine, modeled on the old Russian KGB and China's secret police. A vast network of informers was known to keep the entire population in check, eavesdropping everywhere from tea shops to temples. My guidebook stated baldly, "At some point on your trip (and you'll probably never know when) the authorities will be watching you." The US consulate warned that travelers should expect to have their e-mails and phone conversations monitored and their personal possessions searched.

Yet the officials at immigration and customs simply waved me through. Outside I found a smiling young man from the Motherland Inn 2, the guesthouse I'd booked in advance, wearing a traditional skirt called a *longyi* and holding a placard bearing my name. He led me, along with a small troupe of backpackers, to an ancient green-and-white van baking in the tropical sun, which wheezed improbably to life. As we puttered into town, I stretched my face toward the open window, feeling the breeze hit me like hot breath. Barefoot monks darted across the street. Women carried colorful umbrellas, their faces painted with a chalky yellow sunscreen called *thanaka*. By the side of the road, I spotted a bowl full of dead chickens as limp as feather dusters.

The Motherland Inn 2 proved to be a green, three-story house on the outskirts of town with a cheerful café that spilled onto the sidewalk. After I checked in, I pointed to a map on the wall behind the front desk and asked the girl who did my paperwork if it was possible to travel to the Great Tenasserim River.

"Can, can," she said to my surprise, and told me to apply for permission from the official state-run Myanmar Travels and Tours, giving me directions to walk to the office in the city center.

As I set off downtown, I tried to keep my eyes on the sidewalks, where gaping holes to the sewers below exhaled fetid fumes. But it was difficult not to look up. The leafy avenues of Yangon boast the highest concentration of colonial-era buildings in Southeast Asia. After the British conquered the city in 1885, they rebuilt the capital in grand fin de siècle style. More than a century later, much of the architecture remains untouched, colorful if decaying, like partially erased pencil drawings, smudgy, faded, and streaked with grime.

Across the street from a two-thousand-year-old pagoda marking the center of town, I spotted a storefront with a peeling sign that read in cherry letters MYANMAR TRAVELS AND TOURS. Inside, a dusty fan fluttered a few pamphlets in a mostly empty metal stand.

"Cannot," the woman behind the dark wooden desk said flatly when I asked about visiting the Great Tenasserim River. "Forbidden."

IT SEEMED THERE wasn't much I could do except wait for Heiko, but he wouldn't be arriving for two whole weeks. Fortunately, I had managed to rustle up one local contact to meet in the meantime. Shortly before leaving New York, I'd learned that Kenny the Fish had an exclusive supplier in Yangon named U Tin Win (*U* being an honorific pronounced *ooo**). When I e-mailed him, his reply came quickly: "Dear Emily, You are warmly welcome from Yangon (Rangoon), Myanmar. During staying in Yangon, my son, Mr. Ye Hein Htet, will accompany with you. Sometimes my wife and I will take care of you. My wife and I are over 60 years, so we are not very active."

Sure enough, at 8:59 a.m. the morning after I arrived in the country, a neatly dressed young man with a square face and thick glasses appeared in front of the Motherland Inn 2. This was Ye Hein Htet, who said I could call him Hein. As we drove to his family's farm in the country outside Yangon, he told me how he'd ventured all over Myanmar with his father, "Mr. Tin Win," in search of ornamental fishes—though never up the Great Tenasserim River.

Asphalt eventually gave way to dirt as we passed through an open-air market where women in straw hats hawked vegetables

---

* Burmese people generally don't have surnames, but their names often contain two or three parts. Complicating matters, each name should be preceded by one of dozens of honorifics, depending on your relationship to the speaker.

and flowers. At last we reached a modest green farmhouse encircled by a wall decorated with gold accents and topped with barbed wire. Awaiting us on the stoop were Hein's parents, a bespectacled couple perched side by side: Tin Win in a brown *longyi* with a snowcap of white hair, and his wife, Tin Pyone, who wore a floral batik dress, a black bun pinned at her nape. As they stood to greet me, I saw that neither came up to my shoulder.

Inside, the little house was painted mint green and decorated with gold curtains and dark wooden furniture. On the coffee table someone had set out a tray of orange slices and Cokes next to a neat stack of fish books—one of them a definitive guide to Myanmar fishes by Tin Win himself. Across the room stood a Buddhist shrine, an aquarium containing minuscule golden fish flecked with blue dots, and a large flatscreen television revealing the opening slide of a PowerPoint presentation. "If you can spare your precious time . . . ," Tin Win said, gesturing humbly to the TV.

Before he went into the fish business, Tin Win had been a chemistry professor, and in anticipation of my visit, he'd prepared four informational presentations. Let me summarize what I learned: First, the fishes of Myanmar are much like Tin Win and his family—small, colorful, understated, and (to use one of Tin Win's favorite words) precious. Many are miniature species belonging to the genus *Danio* with tiny spots, speckles, and stripes that you have to squint to see. The fish in Tin Win's aquarium were a new species, which he proudly identified as *Danio tinwini*, named for him in 2009. Also bearing his name is a black-and-white spiny eel called *Mastacembelus tinwini*, described by none other than Ralf Britz. Which leads to the second important point I soon gathered: I was, almost certainly, sitting in the living room of Ralf's secret contact in Myanmar, the one he'd refused to reveal to me.

This became undeniably clear during the presentation entitled "Old timer and recent ichthyologist who interested Myanmar Fish Fauna." After a parade of nineteenth-century naturalists with robust facial hair, I was excited to see a slide featuring Tyson Roberts, whose photo I hadn't been able to find online. (Tin Win called him "Robert Tyson," which was fitting since Tyson, as I later observed, calls Tin Win "Win Tin.") The slide showed a polar bear of a man with stringy gray hair stooped over a microscope, and identified Tyson as "the outstanding ichthyologist" who "started to rejuvenate the study of Myanmar fish fauna after the Second World War."

I was just on the verge of telling Tin Win that I had come to retrieve the Myanmar arowana for Tyson when the slide flipped and a picture of an elfin man with a soul patch in a lime-green T-shirt appeared: Ralf Britz. Actually, Ralf was in a lot of Tin Win's photos: relaxing in a hammock, drinking a beer, wading into a river. Unlike Tyson, he was clearly a close friend of the family's.

As I was to learn, Ralf had not only named a spiny eel after Tin Win and coauthored a paper with him, he'd helped alter the course of Tin Win's life. Tin Win grew up in Mandalay in the 1940s and '50s next to an artist whose goldfish pond mesmerized him. He took up fish-keeping himself in 1972, around the time he met his wife, Tin Pyone, also a chemist. After seven years of marriage, Hein was born. ("Very rare and precious son," Tin Pyone said.) It was hard to eke out an existence as a professor in Myanmar. Tin Win told me his entire salary could buy no more than twelve hundred cups of tea a month (a cup of tea costing a quarter at most). Even on this meager income, as the country was falling apart and the academic system being dismantled, he kept studying fish in his spare time. He couldn't formally describe species because he wasn't trained to do so and didn't have the right equipment. Moreover, like everyone else in Myanmar during those years, he was almost totally isolated from the rest of the world because of the military junta. This meant that no one in the country was in a position to describe the batik arowana or any other native fish. No Burmese ichthyologist had done so since 1963.

In the late nineties, one of Tin Win's students met Ralf by chance in a tea shop in Yangon and referred him to his fish-obsessed professor. When Ralf asked to come see his aquariums, however, Tin Win was terrified. At that time, associating with a foreigner could be dangerous. But he took the risk and a friendship blossomed. Ralf suggested that an untapped market might exist for all of Myanmar's tiny, colorful fishes. In 1998, Tin Win retired from teaching and opened an aquarium fish exporting business. Two years later, he became the country's exclusive supplier to Kenny the Fish.

To conclude his presentations, Tin Win revealed that he knew all about the batik arowana. In the early 2000s, he had first heard about a strange species from the south that looked green when young, but then, as it grew, began to reveal a faint pink pattern eventually darkening to bronze. He had even managed to acquire one, but it was stolen from his house. The authorities caught the thief, who went to prison, but they didn't recover the fish, which Tin Win was sure had been sold to a competitor. He called the species the Chaung Phya Lucky Fish after the stream where he believed it was found. "Maybe I would be the first person to show the fish on the Internet—this is my mistake," he said, explaining that since then the arowana had been smuggled out of the country. "I'm afraid the wild fish will be gone in near future."

He told me Ralf had asked him to get a specimen, but he didn't think he could, because the Great Tenasserim River was insurgent territory.

I confessed that Tyson had asked *me* to get him a specimen.

"I think up to the habitat you cannot go," Tin Win said.

"The government will not allow foreigner to go there—very deep jungle."

NONE OF THIS sounded at all promising, and I spent a lot of time fretting about having roped Heiko into what seemed likely to be a doomed expedition. I kept reassuring myself that he had to know what he was getting into, being no stranger to the challenges of exploration—even here. In the 1980s, when few foreigners were allowed into Burma, he had researched the fishes of the north and written an article that one aquarist told me he'd read so often it disintegrated.

Nevertheless, I felt a pang of guilt every time I heard the repeated refrain that it was impossible to reach the habitat of the batik arowana. With ten days to kill before Heiko's arrival, to keep my mind off the problems ahead, I visited the Shwedagon Pagoda, the most revered Buddhist site in Myanmar, which sits high atop Yangon like a golden crown. I arrived in the early evening as the setting sun lit the 325-foot stupa afire, illuminating the thousands of diamonds and precious gems embedded in its upper tiers. Crowds of devotees circumambulated the terrace, knelt in prayer, or snapped selfies in front of chimerical statues. I watched a monk step gingerly over a cockroach.

Buddhism teaches respect for all life-forms, no matter how lowly, with animals and humans connected through the endless cycle of birth, death, and reincarnation. When the religion spread from India to China in the early first millennium, it brought the practice of liberating captive animals in exchange for good karma. After catching a fish intended for dinner, say, a fisherman might decide to set it free instead.

Around this time, the first records appeared of a wild carp that was usually gray, though every so often a ruddy one popped up like a redhead in a family of brunettes. A thousand years ago, these brightly colored anomalies were considered sacred.

When caught, they were spared from death and released into "ponds of mercy" on the grounds of monasteries. Beginning in the 1200s, the species was bred to be shorter and stouter, longer-finned and twin-tailed—until the goldfish as we know it emerged. For a while, goldfish were exclusively the playthings of royalty. By the 1500s, however, the pet had become so popular that rich and poor alike kept it in ceramic vessels.

Ironically, the ritual of liberating wild creatures had produced the first domesticated pet fish, now circling glass bowls by the millions.

I FELT AS if I were circling a glass bowl myself. The batch of tourists with whom I'd flown in had all moved on, heading north, and the staff of the Motherland Inn 2 seemed to be eyeing me with growing curiosity. Six days before Heiko was due to arrive, looking to escape the scrutiny of the guesthouse, I took a short taxi ride to see three famous albino elephants in a suburb of town. Revered as emblems of purity and power, the animals had appeared up-country about ten years back. Like the sacred carp of old, they were promptly captured and put on display by the powers that be as a kind of symbol of divine right.

The elephants were now housed in a concrete pavilion with an ornate gilded roof. More pink than white, they had pearly eyes and downy blond hair. Chained by their front feet, they rocked back and forth like patients in a mental ward. A sign read IT IS ASSUMED THAT THE NATION WILL BE PEACEFUL, PROSPEROUS, AND TOTALLY FREE FROM ALL DANGERS BECAUSE OF THE WHITE ELEPHANT.

I couldn't bear to stay for long. Afterward, my taxidriver, a young hip-hop artist who had come with me to see the elephants, offered to show me where Myanmar's most famous dissident lived. Aung San Suu Kyi, daughter of the assassinated founding father General Aung San, had been kept under house arrest for the better part of two decades before being freed the previous year. Her home, which had long been off-limits, was now a popular tourist destination, according to the driver.

I began to think otherwise after he dropped me off on a deserted strip of grass between the highway and a stone wall lined with metal spikes. I'd presumed he was getting out too, as he had at the elephant garden, but instead he drove away.

I walked over to a closed iron gate and looked up at a black-and-white photograph of General Aung San flanked by security cameras pointed at my face. There was no way to go inside. All I could do was wait for the driver to come back, which he did after a few minutes.

"Where did you go?" I asked, after I climbed in and he peeled off.

"I had to get gas," he said. "Don't worry. You won't get in trouble." Then he added, "*I* could, but *you* couldn't."

That didn't exactly make me feel better.

I TOLD MYSELF that I was growing paranoid, that no one cared what I was up to. But the next day—five days before Heiko was due to arrive—something strange happened. I came downstairs to get breakfast, and one of the boys at the inn greeted me as usual: "Wazzzzzzzupp?" (He received all the guests in their native speech—"Bonjour" for Paris, "Good morning" for London, "Wazzzzzzzupp?" for New York.) That morning, however, he leaned in and added in a conspiratorial tone, "You went to Aung San Suu Kyi's house!"

I froze. *How could he possibly know?*

When I called Jeff from a phone in the lobby, I told him what had happened and then immediately regretted it, thinking the line could be tapped. He casually advised me to check out the

Wikitravel page for the Motherland Inn 2, saying it had some important information about fleas.

Not only did the website allege that the inn was infested but also that it was "VERY close to the government, as this guesthouse is owned by a son-in-law of a high-ranking army-general! . . . So be careful when dropping your postcards in the mailbox at the reception—maybe it will be read by someone you don't expect. . . ."

I eyed the breakfast boy skeptically as I left to clear my mind, trying to puzzle out how he could've heard where I'd been the previous day. The hot sun beat down on my head, and I started to wilt, stopping for a drink at a tea shop. It was early afternoon by the time I reached the gilded pagoda at the city center, where I decided to pay the entrance fee and have a look inside. The ancient temple was small and more impressive from the exterior. Just as I was getting ready to leave, however, I stopped to inspect a bejeweled fish pendant on one of the altars. A monk approached me, introduced himself, and offered to answer any questions. He was tall, in his twenties, thin and sinewy, with a pointy head.

"What's the significance of this fish?" I asked.

The young monk said it was a symbol of fearlessness. Without fear, he explained, all living beings may swim through the ocean of suffering without danger of drowning. Then he asked me how I liked the Motherland Inn 2.

"How do you know where I'm staying?" I asked, too taken aback to be polite.

"I followed you," he said brightly.

"But I've been walking around for hours."

He nodded, his face maintaining an even expression I couldn't read.

I mumbled that I had to go and skimmed down the stairs toward the exit. Just as I was about to step back into the sunlight, a young woman blocked my path, holding a struggling songbird in my face. At first, I didn't understand what she was doing. "Let it go!" I cried, surprising myself with my anger. Then I realized this was how she made a living: she wanted me to give her money to release the bird for good karma. I quickly shoved a dirty bill into her hand and she opened her fist, the finchlike creature rocketing to the ceiling like a diver coming up for air.

I HAD NO idea whether the monk had followed me out of innocent curiosity or something more sinister. Too many empty days filled with nervous anticipation had whipped me into a tizzy of angst. I decided to leave the Motherland Inn 2 and check into a brand-new Singaporean boutique hotel in the city center. Not only did it feel more private, but it also had air-conditioning and an even bigger luxury—Wi-Fi—so I could communicate with Heiko more easily if necessary. And then, with only two days to go before he was due to arrive, I checked my e-mail, my heart drumming at the sight of a message in my in-box. The subject line was "unfortunate news, terrible."

"Dearest Emily," I read, "I am extreme (terrible) sorry, but I had to cancel my flight to Myanmar today . . ."

CHAPTER THIRTEEN

Beard of the Crocodile

MYANMAR

I was too numb to be heartbroken. Heiko wasn't coming. What's more, I couldn't even be mad at him. He said that his wife Natasha's younger brother was seriously ill, and they'd been fighting all week about what to do. In the end, he couldn't leave under the circumstances.

I had been in Yangon nearly two weeks waiting for Heiko to arrive and pull the kind of superhuman feat he was famous for—though the whole time I'd been losing faith, doubting that even Heiko could reach the Great Tenasserim River. Now that he wasn't coming, I was doubly convinced that I couldn't get there on my own.

Beneath the crushing disappointment, however, I felt a bit of relief. For one thing, I didn't have to worry about luring Heiko onto an ill-conceived expedition. Beyond that, I realized the way I thought about the batik arowana had changed during my time in Yangon. I no longer viewed the fish as entirely "undiscovered." Even if it hadn't been formally described in the literature, it was well-known to a select few in Myanmar. The idea of waltzing down to the Tenasserim and seizing the fish for an American ichthyologist had the distasteful whiff of scientific colonialism. If anyone deserved to determine its fate, I had come to believe that person was Tin Win, who had dedicated so much of his life to studying the otherwise neglected fishes of Myanmar.

Tin Win had, in fact, been checking in with me regularly. When I told him that Heiko had canceled, but that I still wanted to try to find the batik on my own, his son, Hein, offered to accompany me as far as the coastal town of Myeik, the capital of the deep south. Tin Win thought it would be possible to at least see the fish in someone's house there, and I planned to apply (and likely get rejected) for official permission to visit the Great Tenasserim River. Myeik was technically open to foreigners, though they were restricted to the city limits, and my guidebook warned that the frequent cancellation of flights led to "the very real chance of getting stranded down here for days, if not weeks." Nevertheless, I managed to buy two round-trip tickets. On the Tuesday that Heiko and I should have been embarking on our expedition together, I boarded a plane with Hein instead for the two-hour, multileg flight to Myeik.

As we flew over the remote Andaman Sea, I looked down on the turquoise water dotted with some eight hundred isolated islands. For millennia, the vast archipelago had been home to the nomadic Moken, quasi-amphibious hunter-gatherers known as sea gypsies, who roamed about in flotillas, each boat carved from a single tree and shared by a nuclear family. In 2004, the Moken used their knowledge of wind, tides, and animal behavior to anticipate the tsunami that claimed nearly three hundred thousand victims. Though the tribe lived precisely where the giant wave hit hardest, they suffered no known casualties, having retreated to the hills. Since then, the Myanmar and Thai governments had forced many of these people to settle on land in an effort to assimilate them.

Despite the Andaman Sea's reputation as an untouched paradise, the marine biologist Mark Erdmann—discoverer of the Indonesian coelacanth—had told me he'd gone diving there two years back and was deeply disappointed. "We basically arrived like a year too late," he said. "Above water it's still gorgeous. There are spectacular islands and white sand beaches with fantastic tropical forests stretching right down to the beach. But underwater, they let the Chinese in and sold off their fishery resources." He described how at night the lights of the trawler boats looked like a city on the ocean. "Mostly reefs were just trawled and blasted. We had several instances of major bomb fishing right on our heads."

That was the ocean. I was hoping to head in the opposite direction, east into the hills and then down into the Tenasserim River Valley, where the world's longest civil war had been raging since 1948, theoretically halting development and preserving the jungle. Most people in the valley belonged to a minority ethnic group called the Karen, who had been fighting for autonomy ever since Burma won independence from the British. According to the US State Department, Myanmar had a long history of human rights abuses against the Karen. Over the past thirty years, many thousands had been forced to take refuge in sprawling camps across the Thai border.

Just a few months earlier, it had looked as if the war was heating up, with reports of increased skirmishes. Then, mere weeks before my arrival, the Myanmar and Karen armies agreed to an unprecedented cease-fire. For the first time in sixty-four years, a delicate peace had settled over the steamy jungles of the Tenasserim.

To me, it seemed like an incredibly lucky moment to have come looking for the batik arowana. Upon landing at the tiny Myeik Airport, however, I began to realize that getting out of town would not be easy. As we crossed the grassy field from the plane to the exit gate, uniformed security officers in reflective sunglasses stopped us and questioned Hein in Burmese. He gave them the name of his father's friend, a local who now worked in the oil-palm business and had agreed to look out for us. After making a call, the officers waved us through, and I realized Ralf Britz was right—I would've gotten nowhere coming here alone.

The friend had sent an associate to pick us up, a jowly man who pulled up in a white SUV pulsating with club music. He wore Ray-Ban sunglasses and talked into a Bluetooth headset with the rapidity of a stockbroker as we drove into town. (So much for what I'd read about cell phones not working in the south.) Only later did Hein tell me what all the rapid-fire speech was about: the man was trying to figure out how to get me out of the city and up the Great Tenasserim River.

THOUGH I DIDN'T plan to break any rules, there was a heady feeling of possibility in the air and—as I soon learned—not a single arowana to be found in Myeik. This became evident after I met Tin Win's friend, a "big boss" (as Hein put it) clad in a black *longyi* and flip-flops, who readily offered to help in my quest. He took us on a whirlwind, arowana-themed tour of town, which was, by necessity, rather short. It primarily consisted of going to see a tank where there *used* to be an arowana. No one explained this to me. So from my perspective, we'd simply driven to meet a skinny, tattooed man in a sleeveless T, who showed me an empty aquarium beneath a clothesline in his backyard. I told him it was very nice.

According to the big boss, the arowana was known locally as *me chaung mwae* or "hair of the crocodile." He plucked my arm hair to illustrate.

"But crocodiles don't have hair," I said.

Hein explained that the translation was more like "beard of the crocodile," because in the Burmese language no distinction is made between beard hair and head hair—or eyebrows for that matter. I wrote this down scrupulously, only later to realize that crocodiles don't have beards either. Even the fish's name seemed a source of obfuscation and mystery.

By now, I'd attracted a small crowd of curious onlookers. Though Westerners have become a novelty in Myeik, this wasn't always so. Before its modern isolation, the town was once the bustling back door to China, a major port of trade between the kingdoms of Siam (modern-day Thailand) and Europe. That afternoon, as our host finished some work in his office downtown, Hein and I roamed the old British promenade, watching the fishing boats crowd the harbor where the Great Tenasserim River spills into the Andaman Sea. Palm trees swayed in the wind. On a facing island, a colossal Buddha lounged on one elbow.

Peering over the railing, Hein spotted mudskippers using their pectoral fins to shimmy across the mudflats below, just as the progenitor of all land animals probably did some 300 million years ago. He told me that his father had asked him to catch and preserve the species for a museum of Myanmar's fishes he dreamed of opening in Yangon.

In recent years, the practice of collecting specimens for museums has come under hot debate. Critics cite the classic case of the great auk: a nearly three-foot-tall, flightless bird, which ranged from the North Atlantic to northern Spain before being driven to extinction by overzealous museum collectors in the mid-1800s. You might think such examples belong to the past. But something similar happened in modern times to that superstar fish, the coelacanth. Scientists were so excited to discover that the "living fossil" *wasn't* extinct, they nearly rendered it so. Between 1952 and 1992, at least 173 of the fish were captured, most by museums, out of a total population of some 200 to 600 slow-reproducing individuals. The IUCN declared the species critically endangered, and it landed on CITES Appendix I alongside the Asian arowana.

Some critics of scientific collection go so far as to object to the standard practice of preserving "type" specimens to document newly discovered species. In the thrill of the hunt, I'd avoided thinking too much about the uncomfortable fact that collecting a specimen of the batik arowana meant *killing* it. Truthfully, I couldn't imagine myself slaughtering a potentially endangered fish. Yet every ichthyologist I'd spoken with insisted the sacrifice was necessary to formally describe the species. Now that Heiko was out of the picture, I still hadn't resolved what I might do if I actually found the arowana—a scenario that I reassured myself didn't seem likely.

That evening over dinner, however, our host made an unexpected announcement: The next morning, he was planning to take several Malaysian consultants to a new oil-palm plantation outside town. He could send me from there into the Tenasserim Hills, where I'd be able to interview a fisherman who caught the batik for a living. There was just one substantial catch—I might not be able to get out of the car.

This didn't sound like official permission, but I decided I'd take it.

THE NEXT MORNING, I rose in the dark to wait in front of the hotel, my feet sweating from nerves. Shortly after five, an SUV pulled up with three men inside whom I'd met at dinner the night before. The first rays of dawn cast a golden glow as we drove out of town following the Malaysians, a cloud of dust rising in their wake. Thatched huts dotted the roadside. Schoolchildren skipped by in green uniforms. Eventually, we waved good-bye to the other car, which turned off into a plantation, and continued on toward the Tenasserim Hills. As the heat of the morning rose, the air took on a creamy, taut feel, like milk before it boils.

Soon the dirt road began to ascend, and the landscape grew greener and cooler. In two hours, we had reached our destination, Tenasserim Town, a cluster of low buildings with rusting red roofs nestled in the hook of the Great Tenasserim River, where it turns west to the sea. This was the place that Ralf Britz swore I'd never reach—where he said it was possible to find the batik arowana.

I'd been expecting to conduct a stealth interview with a fisherman in the backseat of the car. To my surprise, however, a local official from the Union Solidarity and Development Party—that of the ruling generals—came to greet me personally. A thin, smiling man in a blue ball cap, he led me down a dusty road to a low wooden building containing a spacious room with gleaming floors and wicker furniture.

In one of the chairs sat a tall man in his fifties wearing a sun hat and a fanny pack: the fisherman who specialized in catching the batik. Father of ten and grandfather of another ten, he explained through an interpreter that he had grown up in the Tenasserim and had known about the patterned arowana his entire life. But it had been just like all the other fish in the market—one that people ate—until the mid-2000s when a man named Myint Too came looking for it.

I had not only heard about Myint Too before, I had met him back in Yangon. A competitor of Tin Win's, he claimed he was attempting to farm the batik arowana in a corner of the deep south off-limits to foreigners—though he insisted he hadn't yet sold any of the fish abroad, which couldn't be done legally without a CITES permit.

Yet the fisherman told me that Myint Too had hired him to catch as many arowana as he could each year. At this point, he estimated that he had shipped more than a thousand to Yangon. He said the species used to be plentiful in the wild, even populating the rivulets near town. Now, to find it, he had to travel two days south up a branch known as the Little Tenasserim into thick jungle controlled by Karen rebels.

I asked if it might be possible to visit the area that the fish came from, and the men in the room all laughed. "I think next time you can go there," one of them said. "Because right now they are just signing the treaty."

As for seeing the batik, the fisherman lamented that he didn't have any at the moment—once again, I'd shown up at the wrong time of year. He caught the arowana at high water when it bred. But Heiko had insisted on arranging our trip for dry season.

Disappointed, I thanked the men for their time. The government official walked me outside into the fish-smelling breeze and led me down the street to a wet market where women sat hawking the daily catch. From there, we climbed down concrete steps to the muddy bank, where a spotted pig rifled through trash. I looked up the wide river disappearing into the verdant valley and remembered what Ralf Britz had said: "How many times I dream about going up the Tenasserim. There would be so many unknowns up there."

Just then, one of the men called to us from atop the embankment, announcing we had to go immediately—some other officials were asking questions.

We were back in Myeik by noon.

THOUGH I APPEARED to be out of options to find the batik, our return flight to Yangon wasn't for a few days. So Hein and I passed the time touring the local fish scene. We visited a place that sold varicolored lobsters like cloisonné starships and a processing plant where thousands of pungent purple squid lay drying in the sun.

On our last night in town, our host invited us to a ribbon-cutting ceremony and celebratory banquet on a nearby island where a new fish-processing plant was opening, its products destined for China and Japan. At dusk, we rode a creaky wooden boat across the harbor and ascended a ramp to a posh country club where businessmen lounged outside in wicker chairs, drinks in one hand, cigarettes in the other.

For dinner, Hein and I sat with a group of Singaporean bankers who told us Hillary Clinton's recent visit had unleashed an orgy of investment. Now everyone wanted a piece of the Tenasserim, and there was a big push to clear the forest for oil-palm plantations. I'd noticed an official green sign overlooking the harbor that announced the region MUST BECOME OIL BOWL OF MYANMAR (a play on words since the Irrawaddy Delta is often called the country's "rice bowl").

Shortly after finishing the third course, Hein's phone rang. It was Tin Win, who'd been checking in with us regularly. This time, however, he had big news. A fisherman had just arrived in Myeik with two batik arowana in tow. A local aquarium dealer named U Myeik (apparently named after his hometown) was to take us to retrieve them the next day before our flight. After Hein hung up, he looked shocked. Then he started laughing and could not stop. Finally he told me how his father had ended the call: "Son, if you don't get the fish, don't bother coming home!"

Tin Win planned to send the arowana to Ralf Britz.

THE NEXT MORNING, U Myeik met us at our hotel, an unsmiling young man with shaggy black hair. I rode on the back of his motorbike and Hein followed on a rented one to our destination a few miles away: a thatched hut where a family sold snack foods out the front. Ropes clipped with potato chip bags and candies dangled across the entrance like a beaded curtain. Inside, the small space was crowded with modest furnishings and sacks of rice. In one corner, on a low rusted table, sat a glowing blue cube of a tank so bright it was hard to look at. Inside twirled two arowana far too big for the space.

Yes, these were arowana all right. I'd recognize that pouty, unfortunate countenance anywhere, those flat, reflective eyeballs as charismatic as elevator buttons. At the same time, the fish were different from any I'd seen before, pale green and covered with an intricate bronze, wormlike pattern. They reminded me of the Chin women from the north who tattoo their entire faces, even the eyelids. The squiggles on the fish looked a lot like rounded Burmese script, and I wanted the pattern to *say something*—anything at all. I would've been happy with an accidental ad for Budweiser. But Hein scrutinized the markings closely and pronounced them gibberish.

The fish had been plucked from a remote branch of the Tenasserim and hauled ten hours by boat to Myeik in the care of the fisherman's brother—a skinny, toothy tractor repairman, who now greeted us, beaming, and invited us to sit on mismatched plastic chairs. On the map I'd brought, he showed me his village and where the arowana had been caught, south of the area I'd been. Until recently, he said, fighting had engulfed the region. But twenty days earlier, Karen rebels had descended from the hills sans weapons, just to visit. He handed me a wallet-size photo of his grandfather who received a medal of honor fighting the very same rebels back in the 1940s.

As I scrutinized the sepia, bespectacled face, negotiations commenced. The man asked Hein for $300 US for the two arowana. Hein counter-offered $200. They haggled. Then Hein stepped outside and called his father back home in Yangon, talking in a low voice and nodding gravely. After he hung up, he returned and struck a deal for $250. The sum was a far cry from what the fish might fetch in Tokyo or Shanghai, but it represented a fortune for the fisherman and his family.

The only complication was that Hein didn't want to take the arowana on the plane with us lest he get in trouble. Despite its isolation, Myanmar has been a member of CITES since 1997 (though in practice the treaty is rarely enforced). Even though Hein was not carrying the fish across international borders, he wasn't sure about internal regulations and thought it prudent to have them shipped live to Yangon. The payment would go through U Myeik.

As the men worked out the details, I knelt on the floor to take a closer look at the two arowana, one of which had a bloody scrape on its side where it had probably been injured during capture. Because the pair had been caught together, and because they weren't trying to murder each other (a sign of high affection among arowana), I decided they must be mates.

Poor star-crossed lovers. I felt bad for them, abducted from the dark jungle waters, stuck under this blue alien light beam, only to be sacrificed at the altar of science and bound for the London Natural History Museum. If Ralf Britz declared the batik arowana a new species, one of the pair would be pronounced the "holotype" specimen, the organism fated to forever represent its kind, and preserved for eternity in a cool, dark museum archive.

Examining the fish, at once familiar and strange, like a long-lost Picasso stashed in somebody's attic, I tried to decide whether I thought they truly *were* a new species, or just another variety of the same old Asian arowana. I could only be sure of one thing: the population in the Tenasserim had been isolated for a very long time, but its anonymity was about to end.

This was not at all how I'd pictured finding the batik: an air-conditioned SUV ride that ended with a fleeting glimpse of the Tenasserim River; the holy grail of fish delivered like takeout in a styrofoam container; and an exorbitant sum in Myanmar terms changing hands. As I watched the formerly wild creatures, I felt distinctly uneasy. Technically, I had gotten what I wanted, having tracked down the elusive patterned arowana. Now, however, I had to consider what I might have helped engineer in exchange. Much later, I would feel a sharp pang of recognition when I came across a passage the Victorian aquarist Charles Kingsley wrote in 1855:

> The truth is, the pleasure of finding new species is too great; it is morally dangerous; for it brings with it the temptation to look on the thing found as your own possession all but your

own creation; to pride yourself on it, as if God had not known it for ages since; even to squabble jealously for the right of having it named after you, and of being recorded in the Transactions of I-know-not-what Society as its first discoverer:—as if all the angels in heaven had not been admiring it, long before you were born or thought of.

I wasn't immune to the corrupting powers of the arowana after all. Nor, it turned out, was Tin Win. A little while later, Hein let slip a strange comment. He said that his father, who had originally planned to secure specimens for Ralf Britz, had now had a change of heart: he might keep the fish for himself instead.

OUR BUSINESS DONE, Hein and I boarded the flight back to Yangon. At a layover in the coastal town of Dawei, a gray-haired Burmese man got on the plane and approached me saying, "You must be the one who went to the Tenasserim!"

As the only foreign passenger, I would have been easy to identify, but the incident unnerved me since my brief jaunt to the river was supposed to be hush-hush. It didn't help that a security officer at the Myeik Airport had announced that the authorities were looking for a woman of my description who'd been to Tenasserim Town.

I told the stranger on the plane that he was mistaken. Nevertheless, he introduced himself as the biologist U Tin Than of the World Wildlife Fund (WWF) and asked to speak with me in Yangon.

The next morning, I met him at a new Western-style café next to my hotel, not sure what to expect. "So you saw the Tenasserim River?" he asked again, scooting into the booth and plopping a stack of reading materials on the table, including his own book, *A Guide to the Large Mammals of Myanmar*.

"No," I said uncomfortably. I was still shaken by the security officer's announcement at the airport, and I also felt the need to protect the men who had taken me into the hills.

Tin Than persisted. "You didn't reach Tenasserim Town?"

"Off-limits," I mumbled.

He gave me a quizzical look but moved on, explaining that he had always been interested in studying the river because of its isolated geography. He believed it must be full of endemic fish species that had yet to be discovered.

As we spoke, I learned that Tin Than was originally from Myanmar's western coast, though he had spent the better part of two decades exiled in Thailand, where he landed his job as a field biologist with WWF overseeing conservation in his home country. To investigate the illegal wildlife trade, he had gone undercover in the Tenasserim, living among poachers. Rather than the untouched paradise I'd envisioned, he described a lawless frontier where the few remaining elephants and tigers were disappearing by the day.

When he'd run into me on the plane the previous afternoon, he'd been returning from a trip to assess the environmental impact of a massive, multibillion dollar industrial complex that foreign developers were planning to build in the sleepy southern town of Dawei, 150 miles north of Myeik. The complex would include a deep-sea port, steel and fertilizer plants, an oil refinery, and a coal-fired power plant. A new four- to eight-lane highway would cut through the hills to Bangkok.

The Tenasserim was once again poised to become the back door to China, the route by which Asian goods reach the West. In transforming the region, the Myanmar government hoped to emulate Shenzen in southern China, the onetime sleepy fishing village turned megacity, which produces an estimated 90 percent of the world's electronics. It was this industrial boom in China that had driven the consumer market for pet arowana while rendering 80 percent of its major rivers too polluted to support fish. It seemed the Chinese were hungry for a taste of wildness, so little of which is left.

I could relate. I was not only disappointed to have missed seeing the batik arowana in its natural habitat but also dismayed to realize that none of the jungles I'd visited qualified as true wilderness anymore in the eyes of experts. Not Borneo. Or the Malay Peninsula. Or even the Tenasserim. All these forests had been too severely damaged within a few short decades. According to Heiko only two real tropical wildernesses remained on the planet: one in New Guinea and the other in the Amazon. When he'd backed out of coming to Myanmar, he had again invited me to join him on an expedition into the remote Colombian Amazon. It would be one last chance to find an arowana in the wild—the silver species known as the water monkey. At the moment, however, I was far too exhausted—and frustrated with Heiko—to even think of signing on.

I asked Tin Than about the arowana, but he told me he hadn't heard of it, mammals being his primary area of expertise. Later, however, he started to talk about a strange fish he'd encountered in the distant hills. "I don't know the name," he said, adding that specimens could supposedly sell for tens of thousands of dollars. "They are catching it in the deepest jungle of the Tenasserim forest."

He had come across the fish in tanks on an oil-palm plantation owned by one of the wealthiest and most powerful figures in Myanmar. I'd heard this man was a passionate collector of high-end koi—and a family friend of Myint Too's, the aquarium trader who had monopolized the supply of batik arowana.

When Tin Than showed me pictures he'd taken of the fish, any lingering doubt disappeared: it was indeed the batik. He said workers were going to great lengths, risking their lives to pursue the species in dangerous territory. "They are doing it secretly," he said. "They don't want to tell anyone." Tin Than worried the fish would soon be gone from the wild.

"In previous times, it'd be dangerous, talking frankly like this with you," he said. "But now, no, I think." Then he stared for a long time at the photograph he'd taken of the arowana. "Previously we were as isolated as that fish living at the edge of the rivulets."

THE AROWANA VANISHED. That is, the pair of fish in that glowing blue tank—the ones that Hein had haggled for on behalf of his father—*disappeared.*

Hein and I got back to Yangon on Friday night, and the fish were supposed to follow behind us, arriving in a cargo shipment on Sunday. But they failed to materialize. Not on Monday. Nor on Tuesday.

"Something is wrong with this guy," Tin Win fretted of U Myeik, whose cell phone was now turned off or else rang and rang with no answer. Tin Win asked his friend in the south to look into it. But neither the fisherman nor the fish could be located. The fisherman hadn't been paid yet, so it wasn't as if he made off with the money.

What could have happened? Over lunch at a Chinese restaurant called the Golden Duck, Tin Win suggested that someone powerful might not want the new species described because it would expose the source of the fish and upset a smuggling operation. Had his competitor Myint Too managed to intercept the shipment? Or someone else?

I waited for the fish as I had waited for Heiko. Exhausted. Hot. In a constant state of nervous anxiety. In the midst of this suspense, I received a panicked e-mail from Tyson, asking if I'd gotten my hands on the arowana for him yet. "I have learned that some people are rushing to get specimens," he wrote. *How did he know?*

Out of guilt I put off replying. How could I explain what

had happened? Tyson had set me on the trail of the fish—only to lead his rival Ralf's man right to it.

IN THE END, I had to give up waiting for the arowana to be found. I left Myanmar, the mystery unresolved. But I wasn't even home yet before a baffling e-mail arrived from Heiko. "I have convinced Tyson to describe the *Scleropages* as a new species from Myanmar," he announced, referring to the arowana by its genus, and explaining that he planned to print Tyson's description in the scientific journal *Aqua*, of which he was publisher and managing editor. "But do not tell anyone!"

Heiko asked me for any information I had about where the batik lived. "Tyson was never there but has specimens," he wrote, requesting any high-resolution photographs I'd taken of the species or its habitat. "See these were things I wanted to do with you, and was unable, had to cancel in the last minute."

This was bizarre because the whole point of my going to Myanmar had been to acquire a specimen of the arowana so that Tyson could describe it as a new species. I had failed. So how had Tyson obtained the fish on the other side of the world in South America? It seemed impossible that he could have gotten his hands on Tin Win's vanished specimens, especially in such a short time. Yet Heiko claimed that Tyson's description would be published the following month—a near instantaneous turnaround.

Suddenly, I felt angry: Heiko had abandoned me in one of the most closed countries in the world, where I'd spent nearly a month losing my mind, only to triumph in the face of adversity and find the batik arowana, which was then stolen out from under the country's leading expert on native fishes. Now Heiko had the gall to come in at the end asking for photos and information on habitat. If Tyson really had specimens, why would he need anything like that from me?

I had to wonder—*did* Tyson have the fish?

PART V

INTO THE LAIR

CHAPTER FOURTEEN

The Paradox of Value

NEW YORK → GENEVA

The history of scientific exploration is rife with double-dealing and stolen credit. Ichthyology, in particular, has a long and ignoble tradition of fraud. As early as 1818, John James Audubon tricked fellow naturalist Constantine Rafinesque into formally describing a made-up fish with bullet-proof scales, which he called "the wonder of the Ohio." It took decades for American naturalists to recover their credibility. In 1872, the director of an Australian museum traveling in Queensland was served a peculiar lungfish with a snout like a spatula. He sent a sketch of the dish to an ichthyologist, who named the species *Ompax spatuloides*. This "seventh lungfish" stayed on the books for fifty years until an anonymous report in a Sydney newspaper revealed that it had been fabricated from a platypus beak, a lungfish head, a mullet body, and an eel tail. The responsible party remains unknown.

On my way home from Myanmar, I tried to figure out how Tyson could possibly have obtained the batik. He was in Suriname, after all, a knuckle on the fist of South America. Had someone stolen Tin Win's arowana and managed to ship them clear across the globe right under our noses? Or had Tyson somehow acquired another fish?

Gradually, I grew convinced that a third scenario was far more plausible—this one entirely my fault. I worried that Tyson had been so confident in me, believing I'd succeed in my mission to collect him a specimen, that he'd jumped the gun and promised to describe the fish in Heiko's journal—a fish he did not have and would not be getting.

As soon as I got back to New York, I called Tyson. I was a ball of anxieties. He, at least, was in a glorious mood. "I can hear you fantastic!" he said. "It's as if you're in the same room." Then he added parenthetically, "And I don't have any clothes on. It's Suriname—I'm always naked as soon as I can strip down."

"Sounds like a good life there."

"Well, well, well," he chanted, as if rubbing his palms together. "Okay." Then he fell uncharacteristically silent, waiting for me to fill him in on the outcome of my trip. Instead, I embarked on a long and winding speech to delay as long as possible the inevitable conclusion upon which I ultimately had no choice but to land with a thud: "So I didn't get a specimen, unfortunately."

Tyson was quiet, then said, "Of *anything?* No stingray? Or arowana?" (He'd suggested the stingray as a consolation prize.)

"No. Unfortunately not."

"Wow."

I told him how a single aquarium trader with the possible backing of a powerful tycoon had monopolized the supply of batik arowana from the off-limits Tenasserim River. I even confessed that Tin Win, whose son I'd traveled with, had been trying to acquire a specimen for Tyson's competitor Ralf. "But they couldn't get a fish either," I added quickly, omitting any mention of the vanished arowana.

"That's terrible," Tyson said. "I'm so sad to hear that."

"Really?"

"No! You made my day! Look, Ralf Britz is a three-letter word—a *rat*," he said. "He's a very nasty guy. Period." Tyson told me he suspected—though he couldn't prove—that Ralf had ripped off his work on loaches, discovering a manuscript he'd left in an unlocked closet at the National University of Singapore. Ralf would later say this was complete nonsense, that he wasn't even *in* Singapore at the time. But Tyson maintained that subterfuge had become de rigueur in recent years as a result of all the restrictions on collecting that had multiplied in the latter half of his career. "So the people who work on systematic ichthyology are getting really uptight and really stressed and really competitive. The dishonesty is coming out all over."

It seemed like a good moment to ask about the strange e-mail I'd received from Heiko announcing that Tyson's description of the batik arowana was imminent. I expected to hear it was all a misunderstanding. Instead Tyson fell silent again.

"Okay," he said finally. "Heiko's supposed to keep this quiet. . . . So if he told you something about an article that's going to be in his journal . . . Did he ask you for photos?"

"Yes."

"Well, he shouldn't have!" Tyson cried. "You're potentially a loose cannon."

"I'm just a little confused," I ventured. "You must have a specimen if you're in a position to describe it."

"It's tricky," he said. "I'm not free to divulge anything at all." Then he chuckled. "Wait until it's published."

Unfortunately, I couldn't just sit back idly and wait. Heiko had asked me for photos of the batik arowana, which he wanted to publish in his journal alongside Tyson's description of the species. I had to decide how to respond. On the one hand, I *did* have good photos of the fish. On the other, I didn't understand how Tyson could possibly have obtained a specimen, and I didn't want to implicate myself in a case of scientific fraud, my photos making it look as if he had a fish that he didn't. With considerable horror, I imagined Tin Win in Myanmar seeing my name in the photo credit and believing I'd double-crossed him.

I weighed my options. I could lie and tell Heiko I didn't have photos, though he knew I'd seen the fish. The real problem was that I didn't want to piss him off because—despite my better judgment—I couldn't stop thinking about the Amazon expedition he'd invited me to join at the end of the summer. The largest rain forest on earth loomed in my mind as an Edenic paradise where I could at last find the wild arowana living within true wilderness itself. Heiko's descriptions of trekking through primeval jungle days from civilization held me in thrall. As crazy as it seemed to attempt another adventure with him, I hadn't said no.

This complicated matters, and I agonized over whether to send him my photos of the batik. It was Jeff who finally came up with a solution. At first I told him it was ridiculous, which it most definitely was. But eventually, unable to think of anything better, I watched over his shoulder as he loaded my Myanmar photos onto his computer.

It used to take tremendous patience and skill to capture a clear image of a living fish on film. Today, however, almost anyone with a reasonably good digital camera can do it, and my own shots were pretty decent, revealing the arowana's strange, intricate markings with crisp clarity. Jeff took care of that. Clicking away, he carefully doctored the images, darkening and blurring them, shrinking the files, inserting a large flash on the face of the fish. When he was done, I looked like an exceptionally inept photographer. The hope was that if the photos were bad enough, Heiko wouldn't want to use them.

And it worked. "Thanks," he wrote curtly after he saw them, "but unfortunately these are not printable."

Only later did I recognize just how low I had sunk: to avoid committing fish fraud, *I had committed fish fraud.*

THE AROWANA HAD become a corrupting influence in my life. And I wasn't alone. Most dramatically, during my time reporting on the fish, it had brought down the chief detective of the Indonesian National Police (equivalent to the FBI). The previous March, General Susno Duadji had been sentenced to three and a half years in prison after being found guilty of accepting a bribe related to a dispute with an arowana farm in Sumatra.

Meanwhile, chaos broke out in Malaysia, where the arowana industry was exploding, driving out rice-paddy farmers and flooding the market with fish. Many farms offered investment schemes that sold parent stock and split profits from their offspring. Sometimes these were scams—Ponzi schemes in which the same fish were sold to multiple buyers. Things got ugly. Farmers accused each other of poisoning ponds. Police shot a fish thief dead in broad daylight following a high-speed car chase through crowded streets. Prices skyrocketed and then crashed. In the aftermath, Ng Huan Tong—the Malaysian counterpart to Kenny the Fish, who had spoken so candidly with me about the economic pressure he felt—resigned as managing director after a forensic audit revealed that nearly $30 million US set aside to build ponds was missing. "It was quite bad," one insider told me. "He was refused entry to his shop, his farm, everything."

Closer to home, a prime drama had been playing out in New York. Shortly after I joined Lieutenant Fitzpatrick on his alligator bust in 2009, the arowana smuggler he'd arrested five years earlier at JFK Airport reemerged. Queens resident Simon Chaw—a forty-seven-year-old Chinese-food deliveryman turned natural-gas salesman—was caught, once again, returning from a holiday in his native Malaysia with a suitcase full of arowana packed in water-filled baggies. This time his luggage had been misplaced during a transfer in Hong Kong. When it finally arrived in New York, customs agents noticed it was leaking. The case took more than two years to drag through the courts. Now Chaw was serving a yearlong sentence at the Metropolitan Detention Center in Brooklyn—the same federal prison that once housed John Gotti Jr., head of the Gambino crime family, and al-Qaeda member Najibullah Zazi, mastermind of a plot to blow up the New York City subway system.

When I eventually visited Chaw in prison, I found him to be a sweet-natured aquarist who longed to reunite with his beloved dragon fish, Rosie, whom he suspected was now living in a public aquarium in Texas. (He was not officially allowed to know her whereabouts.) Asked what he found so irresistible about the Asian arowana, he said, "I believe this fish brings peacefulness and good luck." Never mind that his entanglement with the species had landed him behind bars as a convicted felon.

During his arrest, Chaw had complained to a US Fish and Wildlife agent that it was "bullshit" that the Asian arowana was banned in the United States when the fish was mass-produced on farms. While the judge scolded him for this remark at his sentencing, no one in the legal system knew the first thing about the species, which had been described in the press as "a kind of fat goldfish." I had a much fuller picture of things, but even I didn't know what to make of Chaw's objection. The truth was that a basic question still vexed me: Why *was* the fish criminalized in the first place?

IN SEEKING ANSWERS, I traced a river of red tape to the Lacey Act of 1900, the first federal wildlife protection law, still used to prosecute smugglers today. Named for Iowa congressman John Lacey, who warned that the world was becoming "as worthless as a sucked orange," the law was intended to reinforce a slew of state game regulations passed in the late 1800s. By then, American wildlife was in big trouble. Due to overhunting, two iconic, once superabundant animals had all but disappeared: the bison (aka the American buffalo) and the passenger pigeon. The bison was ultimately hauled back from the brink, albeit with cow DNA polluting its genome. But the passenger pigeon was not so lucky. A pretty, pastel species slaughtered en masse for food and sport, it may once have accounted for an astounding 25 to 40 percent of all American birds. The very last one, Martha, died at a zoo in Cincinnati in 1914. Three and a half years later, the last Carolina parakeet, a playful red-green-and-yellow bird hunted for its feathers, died there too. It had been the Eastern United States' only indigenous parrot.

Visionary in its scope, the Lacey Act was amended in the 1930s to ban the transport of wildlife taken in violation not only of state laws but also *any foreign law*. By then, the new American conservation ethic had begun to spread across the globe. As early as 1911, the United States had led Great Britain, Japan, and Russia in signing the North Pacific Fur Seal Convention—the first international treaty protecting wild animals. On the lonely Pribilof Islands in the eastern Bering Sea, the wholesale slaughter of seals, clubbed to death for their silky soft furs, had reduced their numbers from at least 2 million to three hundred thousand. Thanks to the treaty's hunting quotas, the population nearly returned to its historic abundance by the 1950s. The barking, wriggling rookery demonstrated that effective wildlife law needed to be applied across borders.

By the end of World War II, however, Americans could still return from a holiday abroad wearing alligator-skin shoes, carrying a pet macaw, and towing a lion-skin rug, no questions asked. But that was about to change. The newly minted United Nations spawned the IUCN, which in 1963 called for a treaty regulating trade in threatened wildlife. A decade later, representatives from eighty-eight countries descended on Washington, DC, to decide which species to place on three lists stipulating different levels of protection. By 1975, when CITES took effect, the most restrictive list—Appendix I—contained more than five hundred animals and seventysome plants. All these species—including the Asian arowana—were effectively banned from international trade.

A week after I returned from Myanmar, I flew to Geneva for the twenty-sixth meeting of the CITES Animals Committee, held across from the original United Nations headquarters, where the flags of the world flapped in the March wind. Inside the conference center, about 150 scientists from around the globe gathered to make recommendations regarding the world's most vulnerable species. This year, aquatic animals dominated the schedule. There were discussions about listing sharks, dolphins, and corals, about establishing sustainable quotas of sea horses from Southeast Asia, giant clams from the Pacific, and sturgeon from the Caspian Sea.

"When CITES first came along, it was very much about putting a fence around species and protecting them," Tom De Meulenaer, chief scientist of CITES, told me between briefings. A serious, dark-suited Belgian, he had attended the very first meeting of the Animals Committee in 1988, when it was twenty people around a table. At its start, he said, "Everything was about banning everything."

Over time, however, that changed. Experts came to realize that the international market generally poses far less of a threat to animals than local exploitation and habitat loss. Gradually, CITES began to shift from an anticommercial philosophy to supporting sustainable trade. This was largely in response to the greatest conservation success story of the last quarter of the twentieth century: the resurrection of crocodilians. After World War II, crazed demand for crocodile-skin shoes, belts, and bags threatened to wipe the reptiles off the face of the planet. In response, CITES tried banning the trade. Ultimately, however, the solution embraced the opposite approach: endorsing the controlled collection of eggs from the wild to be reared in captivity for skinning and sale. By the mid-1990s, sixteen of the world's twenty-three known species of crocodiles and alligators had made impressive recoveries. Eventually, the illegal trade all but vanished thanks to crocodile ranches.

This triumph led CITES to start certifying Asian arowana farms in the 1990s. More than a decade later, however, when I began searching for the wild fish in Southeast Asia, no one spoke of rebounding populations. Rather, the species seemed to have all but vanished from nature. Perhaps there was a simple explanation: unlike crocodile ranching, which places an economic value on wild populations as the source of eggs, farming arowana is a virtually closed system, providing little incentive to conserve the species in its natural habitat.

For this reason, among others, the commercial breeding of endangered species remains highly controversial. "More often than not, all it does is increase demand," Sabri Zain, director of policy at TRAFFIC International, told me, pointing to the example of tiger farming. While only about thirty-two hundred tigers are left in the wild, an estimated five thousand to six thousand are currently farmed in China for their bones, skins, and meat. The Chinese contend that this will eliminate poaching, but Zain and most conservation biologists disagree. "Even a small, tiny increase in demand would virtually wipe out all tiger populations," he argued.

I was surprised to learn that many experts favor harvesting species from the wild when populations can tolerate it. Yet the animal rights lobbyists crowding the conference found this unconscionable—particularly in the case of pets. The Humane Society International likened wild-caught fish to cut flowers for their high mortality. Other groups objected to keeping animals in captivity altogether. "Personally, I find aquariums to be unfathomably cruel," D. J. Schubert of the Animal Welfare Institute told me. "Even for a goldfish that has to be a horrible life."

Schubert wasn't alone in his thinking. In recent years, a growing animal rights movement across the globe has embraced the once-lowly fish. In Europe, governments have even begun to legislate the matter: In 2004, the city of Monza, Italy, banned keeping goldfish in bowls because their shape limits oxygen and the curvature of the glass gives fish "a distorted view of reality." The Netherlands soon followed suit. In the United Kingdom, the Animal Welfare Act of 2006 outlawed offering goldfish as prizes at fairs, as well as selling live fish to children under age sixteen. This led to the 2010 arrest of a sixty-six-year-old British grandmother who was fined £1,000, placed on curfew, and ordered to wear an ankle bracelet for selling a goldfish to a fourteen-year-old.

While such stories may seem absurd, they could signal a revolution on the horizon. Increasingly, ichthyologists have begun to acknowledge the previously overlooked complexities of the piscine mind. Writing in the journal *Fish and Fisheries*, a British team identified a sea change in the field in 2003:

> Gone (or at least obsolete) is the image of fish as drudging and dim-witted pea-brains, driven largely by "instinct," with what little behavioral flexibility they possess being severely hampered by an infamous "three-second memory." Now, fish are regarded as steeped in social intelligence, pursuing Machiavellian strategies of manipulation, punishment and reconciliation.

Some fish even *play*, according to the American evolutionary biologist Gordon Burghardt, who was initially skeptical while researching his book *The Genesis of Animal Play*, but ended up producing his longest chapter on wallowing carp and garfish leapfrogging over turtles. Among his more compelling examples are elephantfish, which have the largest cerebellums of any fish and seem to enjoy balancing balls on their snouts. Though closely related to these sporty creatures, arowanas had never impressed me as particularly quick-witted. Yet an expert at the New England Aquarium once assured me they exhibit an uncanny intelligence.

I'd long wondered if such animals get bored in aquariums. When I posed this question to Svein Fosså, the former head of Ornamental Fish International, whom I knew from Aquarama, he pointed out the inconsistency of fretting over the ennui of an arowana when many, many more fish get squeezed half to death in trawlers and left suffocating on the decks of ships. Given the choice, he imagined a fish would opt to be a pet. "If they could think, which I somehow doubt that they can," he added. "I believe many animals, including fish, are sentient beings—but what does *sentient* mean?"

In the year I'd known Fosså, a tall Norwegian with round blue eyes, circular glasses, and apple cheeks, I had come to think of him as the resident philosopher of all things pet fish (though he himself preferred "companion animal" to "pet"). Having attended CITES meetings for the aquarium industry since 1999, he told me that in the early days he was naive about how the convention worked: "I thought it was about science, rational thinking. How wrong I was!"

In Fosså's view, it was all politics. As an example he cited the 2010 proposal to list the Atlantic bluefin tuna on Appendix I. A popular sushi fish whose numbers have plummeted more than 80 percent since the industrial fishing era began, the species is clearly threatened by international trade.* Yet powerful lobbyists, led by Japan, succeeded in blocking the listing. Meanwhile,

---

* The bluefin tuna—not the Asian arowana or the koi—is probably the world's most expensive fish. In 2013, a single 489-pound specimen sold for $1.8 million.

an entire genus of Central American tree frogs, half of which are common, landed on Appendix II because no one cared enough to object. "They got on as a whole bulk simply because there had been so few other listings that everyone thought, 'Well, we have to list *something*,'" Fosså told me in Geneva.

"Even the people proposing it said they're not rare," added his friend Jim Collins, the only other lobbyist in attendance from the pet industry. A British reptile enthusiast who owned one of the world's largest rattlesnake collections and whose best friend had died the previous year from a king cobra bite, Collins glared across the room at the animal rights activists, whom he called "lunatic people."

"They wouldn't use such nice words for us," countered Fosså.

Both men resented the activists for seeking to ban all trade in wildlife. "You could ask them if there should be trade in locusts, and they'd say no," Collins complained, arguing that listing animals without cause was counterproductive, since people actually collect CITES species for the thrill.

Fosså agreed. "As soon as you start to scream about something being very rare, there are always people out there who want to have it."

I had heard this argument before—not only from Heiko but also within academic circles. A few years earlier, a team led by the French biologist Franck Courchamp sought to quantify how humans value rarity by measuring the hassle visitors to a zoo in Paris would endure to see certain animals—how much they'd pay, how many stairs they'd walk up, and whether they'd dart through a sprinkler. The researchers found that people spent about twice as long looking for a frog or gecko marked rare as they did exactly the same frog or gecko marked common.

Standard economic theory has long held that trade alone is unlikely to drive a species to extinction because of the escalating cost of finding the last individuals of a declining population. In other words, at some point you have to shell out more to catch the last arowana in the swamp than the fish can fetch on the market. Courchamp wasn't convinced. In 2006, he and others proposed a new model, integrating a basic facet of human behavior well-known to economists since Adam Smith—the "paradox of value," otherwise known as the "water and diamonds paradox." While water has tremendous practical value, it's worth nothing in exchange. Rare animals are the opposite, like diamonds. Their exaggerated value fuels their disproportionate exploitation, rendering them ever rarer and thus more desirable until they're sucked into an extinction vortex.

When I asked Fosså if he thought declaring the Asian arowana one of the world's most endangered fish had been a self-fulfilling prophecy—the equivalent of releasing a limited edition—he considered the question carefully. "It's difficult to answer," he said, "but at the very least the interest in it escalated tremendously because it was put on CITES—people discovered it."

As for how the species landed on Appendix I in the first place, I hadn't been able to shed much light on the matter. The few people who recalled the early days of the convention characterized the proceedings as a bit haphazard. Animals were listed en masse without any real debate or scientific evidence. The people who made decisions often did so in conversation, leaving behind little documentation. No one seemed hopeful that I'd be able to dig up much history.

NEVERTHELESS, I SKIPPED a meeting on sea cucumbers and hopped a bus across town to the CITES Secretariat archives. Most of the early files had been tossed, and there was just one bookshelf of binders. Still, the task was daunting as I sat at a conference table and began flipping through the mountain of yellowed pages.

I already knew that in the mid-1960s the IUCN had created a group dedicated to saving freshwater fish. In Singapore, I'd tracked down Eric Alfred, the former curator of zoology at the National Museum, who'd been charged with identifying species under threat in Southeast Asia. He'd traveled up and down the Malay Peninsula, where he'd grown up in the 1930s, when it was still "a big wild place," asking fishermen which species were getting harder to come by. Ultimately, Alfred recommended two be added to the IUCN Red List of Threatened Species, both slow-reproducing food fishes popular among anglers: the massive Jullien's golden carp and the Asian arowana. The pair then appeared on CITES Appendix I in the 1970s. The question was—why? Out of all the fish in the world, why add these to a treaty restricting international trade when the problems they faced were local—overfishing and habitat loss?

By the time I met Alfred in 2011, he was in his eighties and had left the fish world behind three decades earlier. The pristine streams of his youth were now empty; in their place stood endless oil-palm plantations. He confessed he had no idea that the Asian arowana had gone on to become such a fashionable, high-priced pet. As we sat in his front garden on a quiet hill overlooking downtown Singapore, he told me what the years had taught him: "As far as conservation is concerned, we should protect the environment, and not the animals. Let the species take care of themselves."

I did finally dig up the original listing for the Asian arowana in that mountain of papers. It was marked D—for "delete." In other words, the fish was meant to be removed from the draft for a lack of evidence supporting its inclusion. For some reason, it remained. No one could say why—or what unintended consequences its fame may have wrought.

CHAPTER FIFTEEN

Fish Meets World

NEW YORK → STOCKHOLM

A nonymity may be a safer bet for a fish than fame. But I had scarcely been back from Geneva a month when Tyson's description of the batik arowana appeared in the April 2012 issue of Heiko's journal *Aqua*. The aquarium world buzzed with the news. "*Scleropages inscriptus* is a new species of arowana!" a popular blog proclaimed, noting that it was "the last kind of fish we expected to still exist undiscovered since arowanas hold a near god-like position in Asian aquarium culture." One commenter declared ominously, "This fish will need its own bodyguard."

Tyson had named the species *inscriptus* for its complex maze-like markings and placed it in the genus *Scleropages* alongside the Asian and Australian arowanas. As for the mysterious type specimen, the paper claimed that a pair of fish had been obtained dead from a vendor in Myeik and then somehow transported to Thailand, where they were now deposited in a natural history museum. I stared at the photograph of a lifeless fish sprawled on its side, mouth agape, its one visible eye cloudy. It was roughly the same size, with the same squiggly pattern, as the two arowana that had been stolen from Tin Win. Sans living vitality, however, the fish looked totally different—flatter, grayer. There was no mention of who acquired it.

"The guy's anonymous and I'm not going to reveal that," Tyson said when I called to congratulate him. I caught him as he was packing up to leave his apartment in Suriname, his health having improved enough to allow him to fly as far as Stockholm, where he planned to spend the summer. "I would be doing my friend very wrong. He's in the aquarium trade, and let me just say there's more than meets the eye."

What I really wanted to know was how Tyson had described physical specimens halfway across the world. To pin down the arowana as a new species, didn't he and the fish need to be in the same room? Asking him this felt like tickling the nose of a bear with a cattail. He swatted me away and excused himself to check on a steak he thought he'd left on the stove—only to remember that he ate it. "Last time somebody called I had spaghetti on," he said. "That burned so fast it wasn't even funny."

Oddly enough, the person who came to Tyson's defense was Ralf Britz, who e-mailed me Tyson's species description along with a line in Latin: *Alea jacta est*, "the die has been cast." I pictured Ralf swiveling around in a chair stroking a catfish. Yet when I reached him in London, he surprised me by allowing that you *don't* have to be in the same room as the specimen you're describing. "Of course you would *want* to," he said. "You would want to look at the fish yourself and not take for granted what other people tell you." But it wasn't absolutely necessary.

So this was not a case of ichthyological fraud, as I'd feared. It seemed Tyson's anonymous collaborator had simply sent him measurements and photographs of the specimens, which was fine, though not exactly ideal. As for whether the fish truly constituted a new species, Ralf agreed that it probably did, though he objected to Tyson's making the call based exclusively on coloration. If the batik arowana was its own species, why weren't the Super Red, golden, silver, and green also distinct species, as that scandal-plagued Frenchman Pouyaud argued?

Part of the problem, I had come to appreciate, stemmed from the surprising lack of consensus over what a species is. Though the concept is as fundamental to biology as elements to chemistry and particles to physics, biologists have never been able to agree on a definition. Darwin, for one, declared the effort futile. "It is really laughable to see what different ideas are prominent in various naturalists' minds, when they speak of 'species,'" he wrote in 1856. "It all comes, I believe, from trying to define the indefinable."

Today more than twenty different species concepts are floating about. The one most of us still learn in high school comes from ornithologist Ernst Mayr, who in 1940 defined species as populations that interbreed. This is the "biological species concept," and many scientists still believe it's the best we've got. Yet detractors dismiss it as "a textbook legend," pointing to the prevalence of hybrids in nature—ligers and tigons (crosses between a lion and a tiger), zonkeys (zebra and donkey), and grolar bears (grizzly and polar bears), to name just a few. More critically, lots of organisms, such as bacteria, don't reproduce sexually, though we divide them into species anyway.

To remedy these problems, the 1960s saw the rise of an alternate definition, the "phylogenetic species concept," which ignores sex altogether, instead relying on descent from a common ancestor. In theory, this makes a great deal of sense—related organisms share traits because they share the same evolutionary history. In practice, however, it upends the intuitive categories into which we traditionally divide the natural world. The class of reptiles, for example, must now include birds, which evolved from dinosaurs. And fish? It's impossible to prune the tree of life and come away with a branch that *just* includes fish, since amphibians, reptiles, birds, and mammals all represent twigs stemming off the same lineage. This led Stephen Jay Gould to regretfully report in 1981 that "there is surely no such thing as a fish."

Gould was being facetious, pointing out how radically the new thinking challenged traditional classification schemes going back to Aristotle. Though he wouldn't have said so, the Greek father of biology relied on the age-old "morphological species concept," which the American geneticist George Harrison Shull defined nicely in 1923, writing that species should be identifiable based on "simple gross observation such as any intelligent person can make with the aid only, let us say, of a good hand-lens." That was basically how Tyson argued the case for the batik.

As much as Ralf would have liked to describe the new arowana himself, he told me he was glad the fish was at last on the books. "Now it's an entity. Before it was not," he said, pointing out that once named the species could be added to databases such as the IUCN Red List. "With that comes a lot of necessary publicity for the fish, which might help in getting to know more about it and maybe even saving it from possible extinction."

Such publicity, however, also has a flip side. In 2006, a team of biologists from the United States and Laos warned in the pages of *Science* that scientific description can imperil species by advertising "novelties" for hobbyists and driving new markets, serving as a veritable "treasure map for commercial collectors." The authors likened the dilemma to that facing researchers in fields with applications to bioterrorism. You wouldn't publish a how-to guide for mailing out anthrax. Maybe you also shouldn't publish a treasure map to a fish.

Just a few weeks after Tyson introduced the batik to the world, Kenny the Fish's head research scientist, Alex Chang, revealed that his company had obtained two specimens. When Alex showed me a grainy cell phone video of the prized acquisitions, I thought *they* looked like the ones I'd seen in Myanmar. I had to wonder if Tin Win—who was, after all, Myanmar's exclusive supplier to Kenny—had actually retrieved the arowana from the airport without telling me and shipped them out of the country to his famous client.

Somehow I doubted it. Alex said the fish came from a supplier in Malaysia, which I found plausible. Either way, however, they had to have been smuggled out of Myanmar—almost certainly from the wild.*

DESPITE THE RISKS of describing new species, I was inclined to agree with Ralf that we cannot protect what we do not understand. What's more, I had come to appreciate just how much remains to be known about fish. When I first set out to report on the Asian arowana, I figured I would start by finding the researcher studying the species in the wild—only to realize no such person exists. The popular illusion that modern science has the entire living world covered, that there is an expert analyzing every crevice, is far from true.

Up to the middle of the twentieth century, biologists usually specialized in one animal group—fish or birds or earthworms, for example. Then in the 1950s and '60s, following the discovery of the helix structure of DNA, the molecular revolution ran through the field "like a flash flood," in the words of E. O. Wilson, monopolizing funding. The ranks of molecular biologists racing to decode the genome swelled, while experts on groups of organisms were gradually pushed out, their work dismissed as Victorian science. "Nowadays it's hard to find people who are solely ichthyologists," Tan Heok Hui, the fish man at the National University of Singapore, told me. "Some people believe it's a bit passé."

* While it's possible to argue that a species new to science isn't subject to trade restrictions, since its name doesn't yet appear on any protected lists, CITES quickly issued a notice that the batik arowana was to be considered one and the same as the Asian arowana for purposes of international trade.

In Tyson, I had found one of the last great traditional biologists trained in the old science of classification called systematics. As one young colleague told me, "He's probably identified more species of fish than anybody else alive." Eager to finally meet the man himself, I flew to Stockholm a month after the publication of Tyson's paper and made my way to a side entrance of the Swedish Museum of Natural History, where he was spending the summer. A large figure wearing a red-striped dress shirt and gray pants held up by black suspenders lumbered down the stairs to let me in. He had a craggy nose, unkempt gray hair, and eyelids that drooped over pale blue eyes.

"Nice of you to come!" he said brightly, as he led me down a hall to a spacious laboratory with white cabinets and large windows overlooking a green lawn. Practically every available surface was stacked with glass jars of various sizes filled with amber liquid in which floated the strange, ghastly forms of long-dead fishes. "All the jars you see are collections, mostly unsorted," Tyson said. "I'm in the process of sorting them."

Over the course of fifty years, Tyson had left a trail of pickled fishes behind him as he traipsed across the globe like some pagan god of ichthyology holding a fishnet in place of a trident. He had collected all the specimens crowding the room from Africa in the 1960s and '70s, back when he was a graduate student at Stanford and then an assistant professor at Harvard. When he left Harvard in 1975, after failing to get tenure in an age when his specialty was no longer in demand, he fought to take the collection with him to the California Academy of Sciences, where he stayed five years. Since the 1980s, he had been a research associate at the Smithsonian Tropical Research Institute in Panama, but it was a loose affiliation. Really, he was an itinerant, moving from museum to university to museum. "He travels like crazy," a longtime colleague told me. "Nobody knows where he gets his money. He lives on a real shoestring."

Now Tyson heaved himself into a chair at his desk to show me a manuscript he was proofing on oarfish, which he described as his masterpiece—or *one* of his masterpieces. Oarfish are enormous, near mythical oceanic fish of the genus *Regalecus*, meaning kingly, for the red, crownlike crests atop their heads. They're the longest bony fish on earth, documented at twenty-six feet—the height of a two- or three-story building. Ancient mariners likely mistook them for sea serpents, though the fish are rarely seen unless they leave the ocean, which happens from time to time. "They swim out of the water onto the beach, effectively committing suicide," Tyson explained, showing me a photograph of some thirty US marines hoisting up the carcass of an oarfish in California. Once the fish decides to come ashore, it's impossible to change its mind. Lasso the creature and drag it back into the ocean, and it just charges toward land again. Solitary, recalcitrant, drifting across the oceans of the world—the species reminded me of Tyson.

"I see I've got some fish I didn't finish sorting," he said, looking across the room. He snapped his suspenders, hoisted himself up from his chair, and trudged over to the windowsill, bending in half to inspect a cluster of specimens. "Let's see if I can find this thing I saw the other day. . . . Ah, yeah, here it is!"

He picked up a tall jar, leaned in to unscrew the lid, and pulled out a dripping wet fish about the length of a shoe with a snout like a pair of pliers. It was an elephantfish, one of those brilliant creatures with enormous brains that researchers found will play ball in their tanks. Yet Tyson believed this one was new, different from any he'd ever seen.

He tried to pry open its jaw and then handed me the fish. "See if you can get that out," he said, indicating a slip of paper stuffed in its mouth like the fortune in a really disgusting fortune cookie. The fish was corpse cold, stiffer than I expected, and smelled like chemicals that would give me cancer. My pinkie was halfway down its throat when I spotted a pair of tweezers and altered course. The slip of paper said the fish was *Mormyrops nigricans*. Tyson didn't think this was right. He tromped back to his computer and looked up the species. "Not the same, huh?" he said, looking between the fish in my hands and the one on the screen. "Or is it the same?"

The two fish looked pretty close to me. But Tyson leaned in and started counting the scales. "Look how large they are!" he said, poking the fish. "See! See what I'm talking about!"

I did not. I was too busy thinking about how the fish went into the jar in 1973, when Tyson was a young hotshot professor at Harvard, carrying out a survey of the Congo River in Zaire for the National Geographic Society. Then some forty years later it came out of the jar, and—poof—Zaire no longer existed (it was now the Democratic Republic of the Congo) and Tyson was a nearly seventy-two-year-old man of no fixed abode with high blood pressure, chronic thrombosis, arthritis, and foot pains, which he blamed on abusing his feet during fieldwork. Obesity had crept up on him in middle age. Sleep apnea damned him to perpetual exhaustion. If he put the fish back in the jar, I wondered, would it ever come out again?

"Does it bring back memories going through your fish?" I asked.

"Tremendous memories," Tyson said. He told me about the lake where he caught the elephantfish, Lake Mai-Ndombe, which has water as black as squid ink and is big enough to be seen from the moon. He discovered a new fish there, the female of which outshone the male—a rarity in the animal world—and named it *transvestitus*.

As a boy growing up in Los Angeles in the 1940s, Tyson got his start collecting birds' nests with eggs in them until his mother told him that if he continued, there would be no more birds. So he switched to rocks and minerals and fossils and fungi and grasshoppers and butterflies. Eventually, he began netting crawdads in the Los Angeles River, the banks of which were still semi-wild in those days. Soon he developed a passion for fish, and at Stanford University, spent hours upon hours studying its world-class collection.

According to Tyson, he decided as a child that he would never marry or have children because he knew he wanted to be a tropical biologist and travel all his life. Because of his singular obsession, we now know a great deal more about the fishes of Africa, Southeast Asia, New Guinea, and South America than we otherwise would. In many instances, the aquatic ecosystems he studied have since been destroyed.

Yet Tyson lived through an age when traditional biology came to be looked down on as mere stamp collecting. Natural-history collections no longer receive funding for proper mainte-nance. In some cases, they're even being tossed out. Despite the recent debate over the merit of faunal collecting, such libraries of life are the foundation upon which biology was built. At the Swedish Museum of Natural History, we were literally standing atop a basement archive filled with specimens Linnaeus himself preserved some 250 years ago.

I followed Tyson down the hall to the regular library, where he wanted to look up the curious fish from the jar in the *Catalogue of the Fresh-Water Fishes of Africa* from 1909. He found three leatherbound volumes and heaved them onto a table. The first book fell open to a drawing of an elephantfish that had the profile of an old man with an underbite. "I've always liked this one," he said. "Look at that face."

Then he flipped several pages and spotted a slim, tiger-striped fish with sharp teeth and a forked tail. "This is the one species in the Congo that I was never able to collect. Never got it . . . never ever got it . . ." He repeated the phrase a half dozen times, shaking his head. Never mind that he had caught practically every other known species in the Congo. This was the fish that got away.

I could relate. I wondered if fifty years hence I would still be haunted by the wild arowana that had so far eluded me. I also wondered if anyone living a half century from now would know as much about fish as Tyson. When the current generation dies, no experts will remain for many groups of organisms, some of them quite large. Species may disappear without anyone noticing.

"These people who are molecular biologists are essentially ignorant," Tyson complained. "They don't really understand the relationships between things." Reducing life to genes incensed him. Recently, American companies had begun patenting naturally occurring segments of DNA. "To me this is an obscenity," he said. "I don't believe in God. I'm an atheist. But it's un-*godly*!"

The following year, in 2013, the US Supreme Court would rule against such patents. Nevertheless, the world was moving toward a more radical commodification of life. In 2010, the Convention on Biological Diversity had adopted the Nagoya Protocol, which created a legal framework for enforcing states' ownership of their biological resources. In theory, this could mean, for example, that Southeast Asian farmers raising fish from the Amazon may eventually have to share their profits with the South American countries the fish originally came from. In 2011, Malaysia had patented its prized golden arowana. Now the Indonesians were pushing to do the same with their own Super Red.

As I sat across from Tyson in the library, watching him stew about the rush to stake claims to the organic products of evolution and natural selection, his face grew red and his fists clenched. Suddenly he exploded. "*Shoot them!* Kill every son of a bitch down to the last person involved!" he cried. "No! Never, never, never, never, never!"

Only later did I realize where I'd heard this line before. Tyson, like some mad king of ichthyology, had uttered the dying words of King Lear.

I NEVER FIGURED out how Tyson got the batik arowana, though I have my ideas. I think the fish came from an aquarium importer in Thailand who got a supplier to slip them out of Myanmar. According to Tyson's paper, the two specimens he described were obtained in 2011—the year before I even set foot in Yangon. Maybe Tyson didn't know about the fish until later. Or maybe the fact that he encouraged me to pursue a third specimen when he already had two demonstrates how avaricious collecting can be.

Either way, the batik arowana quickly became old news. Once described, it didn't seem to interest Tyson anymore. I'd moved on too: my sights were now set on another fish, the Amazonian silver arowana, *Osteoglossum bicirrhosum*, which Tyson had encouraged me to write about the first time we spoke. Back then I'd brushed off the suggestion. The species isn't listed as endangered, so it's legal to bring into the United States, where small ones sell for as little as $30 to $50 apiece. Evidently, I valued rarity as much as the next person—I wasn't interested in a supposedly common fish.

According to Tyson, however, the silver was the true *arowana*, a name that comes from the indigenous language group Tupí-Guaraní, from which we also get words such as *jaguar* and *tapioca*. Together with its African cousin, the silver arowana was the first to be formally described in 1829, and the first to enter the aquarium trade in the mid-twentieth century. Compared to the various color morphs in Asia and Australia, it looks a bit like an unadorned template—the model as initially conceived. Slightly larger than the others, with an even more flexible and snakelike spine, the species is the best jumper of the bunch, which has earned it the nickname water monkey. A nearly vertical drawbridge of a mouth gives the fish a sort of square-jawed, robotic look. Softening this impression is its tapering tail, which appears to be adorned with a single doily-like ruffle.

Despite its comparatively low price, Tyson told me the silver arowana was a highly profitable aquarium species in the Amazon. He described how fishermen ventured into the swamp each year during breeding season and collected the young from the fathers' mouths. Recently, however, I'd heard that the annual catch had fallen about 70 percent. The shortage was attributed to flooding, which restricted the access of fishermen and dispersed the arowana deep in the submerged forest. Yet I wondered if more than the weather was to blame.

Over the past few years, China had stuck a straw into the Amazon and was slurping out silver arowana by the millions, marketing them as "starter" pets—dragon fish within reach. On my way to Borneo the previous fall, I had stopped in Guangzhou, where I'd met with a former employee of Kenny the Fish's, Lin Jieming, who had gone into business on his own selling silvers. Lin took me to the facility where he quarantined the thirty thousand wild arowana he imported every month before distributing them throughout the country. The Chinese, he said, bought more than a million of the fish each year—though there was demand for twice as many. His suppliers reported that it was getting more and more difficult to find the species.

Was the silver too dwindling to eventual rarity? I hoped to answer the question by joining Heiko in the Amazon. At least one ichthyologist did not think this was a good idea. "Look, just take my advice—don't make any field trips with Heiko *anywhere*, because he'll let you down," he warned me. "I wouldn't go with him for all the tea in China."

I wasn't worried that Heiko would stand me up again because, unlike our misadventure in Myanmar, this expedition was not my idea but *his*. What did concern me was his apparent lack of regard for safety. He told me he had never in his life made a decision based on fear, which I was beginning to think could be true. The year we met, for example, he had casually invited me to join him on a collecting expedition to Colombia's Caquetá Valley—which I later read was the largest cocaine-processing region in the world, controlled by drug traffickers and guerrillas. That was not a detail he mentioned.

In recent years, Heiko had been entering the Amazon via Colombia rather than Brazil. He said this was because the Colombian Amazon contained the least explored tracts of virgin rain forest. Others said it was because he wasn't welcome in Brazil since being arrested for alleged biopiracy and jailed in Manaus. "I would never, ever, in my life go on an expedition with him, even if someone paid me," one aquarium world insider told me, "as I would be one hundred percent sure I would be breaking laws."

All this made me considerably uneasy. So after returning to New York from Stockholm, before committing to join the two-week expedition at the end of the summer, I dutifully set about investigating the obscure affluent Heiko planned to ascend called the Calderón. The Amazon is not a single river but rather a vast capillary network, containing some fifty thousand navigable miles. Two of its tributaries—the Madeira and the Negro—rank among the ten largest rivers in the world. Generally, however, only locals and a few academics know the names of even the Amazon's biggest waterways.

Such was the case with the Calderón. Not much information was available, but the one government report I dug up said that the "drug traffic bonanza" had ended in the area in the 1990s. In 1998, an apocalyptic sect called León de Judá moved in. The group worshipped a diminutive, bad-tempered former cobbler named Ezequiel Gamonal, who predicted the world would end in 2000. Come the turn of the millennium, the world did in fact end—for Gamonal, who died that year at eighty-two. After his death, the community stayed put, continuing to wear Old Testament clothing modeled on Cecil B. DeMille's Hollywood epic *The Ten Commandments*. I examined pictures of men in togas with bushy whiskers and felt enormous relief. Compared to gun-toting narco gangs, I could handle a postapocalyptic cult. What's more, the report confirmed the presence of arowana—though it raised concerns that the population had been seriously diminished by the use of poison for fishing.

Reassured, I made plans to meet Heiko in Bogotá in late August. Then, two weeks before I was due to depart, I realized with some horror that I'd misunderstood our itinerary. The Calderón was the *second* river we'd be visiting, time permitting. The first and most important was the Putumayo, a major tributary of the Amazon, which marks the southern border of Colombia with Ecuador and Peru before flowing into Brazil, where it's called the Içá. I rushed to my recently purchased *Smithsonian Atlas of the Amazon* and read, "The Putumayo-Içá is one of the most isolated valleys in the Amazon Basin." So far, so good.

The text went on to note that few scientific studies had been conducted in the region due to its remoteness—as well as *all its drug traffickers and guerrillas*. Specifically, the Putumayo was a stronghold of the FARC, the Spanish acronym for the Revolutionary Armed Forces of Colombia, the Marxist-Leninist group notorious for kidnapping anyone within reach and holding them in the jungle. "It's a hot spot, and I'm not talking about the temperature," an American affiliated with the US State Department replied to my frantic inquiry. "It's one of the last holdouts for the guerrilla groups. Americans have been known to disappear there for years. . . . I'm sorry if I'm freaking you out, but that is one place I wouldn't go to."

I found a scholar writing a book on the Putumayo only to learn she was too afraid to travel to the river herself. I called an anthropologist studying an adjacent region in Peru who told me he'd stayed clear of the Colombian side of the border since a fellow University of Wisconsin student was abducted there in the 1980s. Things were looking pretty bleak when I came across a new project devoted to conserving biodiversity in the Putumayo Basin run by the World Wildlife Fund (WWF). When I reached its coordinator Camilo Ortega in Bogotá, I learned he had just returned from a town on the Upper Putumayo, where he attended a festival celebrating none other than the arowana. He sent me photos of colossal papier-mâché fish being marched around like dragons in a Chinese New Year's parade.

This seemed highly promising. But when I asked Ortega about working in the area, he said, "For sure it is difficult," and then admitted, with an embarrassed laugh, that his team had been "something like kidnapped," which he pronounced kid-náp-ped, earlier that year. The victims had only been held a few days. After that, however, WWF had stopped sending their people into the region.

The Putumayo, however, is not a small river. A thousand miles long, it passes through four countries. Trying to characterize its entire length was a bit like inquiring about the political situation from Tehran to Tel Aviv. When I gave Ortega the name of the village where a small plane was supposed to deposit us, he seemed surprised. "It's very, very difficult to get there," he said, doubting guerrillas would be in such a remote location. He didn't think anybody would be there at all.

In the end, I decided to trust Heiko. My anxieties aside, he had been exploring the Amazon since the 1950s. "He is half-born there," his friend Holger Windeløv reassured me, when I reached him at home in Denmark in a fit of worry just two days before my departure. "He has been there since childhood, in the jungle."

Windeløv was a retired aquatic-plant importer, who used to buy from Heiko's mother. I called him because he was one of the only people I knew who'd actually been on an expedition with Heiko. In 1996, the two men descended the Iténez River along Brazil's western border with Bolivia, retracing the path that Amanda Bleher had taken with Heiko and her other children in 1954. Heiko had told me about the trip—that the jungle he remembered had disappeared, having been cleared for cattle ranches, and the formerly barefoot Indians now wore rubber boots and watched TV. By Windeløv's account, however, the expedition was the adventure of a lifetime. For more than a week, they passed few people. Heiko looked for fish, Windeløv for aquatic plants. He told me Heiko's determination knew no bounds. He kept the boat out past midnight, diving in the black water.

"Aren't there things in there that are dangerous?" I asked.

"Yes, but danger Heiko don't think about," Windeløv said. "He is frightless. . . . Is that a word?"

For some reason, I told him it was.

Windeløv had just one warning: "Don't think you can decide what you're going to do there—*he* makes the decisions. If you're with him, you have to follow him. Or he'll leave you."

CHAPTER SIXTEEN

Plan C

COLOMBIA

Though I hadn't seen it coming, it was predictable—perhaps even inevitable—that I would end up in the Amazon. Most fish stories do. South America boasts more species of fish than any other continent because the largest river on earth runs through it. What begins as a trickle from a spring in the Andes, the mountains rising high and dry above the Pacific Coast, gathers energy as it flows eastward, ultimately spilling into the Atlantic Ocean with such force that it's possible to taste freshwater more than a hundred miles out to sea. This massive river has a correspondingly massive basin, draining an area nearly the size of the continental United States, which supports the world's largest rain forest and the mother lode of earth's biodiversity. More than three thousand species of fish are known to live there.

By pursuing one of them—the silver arowana—I was setting the bar lower than I had on my previous adventures. This time I wasn't chasing the highly coveted and extremely rare Super Red in deepest Borneo, which no one expected me to find. Nor was I seeking to discover the holy grail of new fish species in an off-limits war zone, as in Myanmar. Instead I had moved on to Plan C: enlisting the greatest living freshwater explorer to help me track down a fish he seemed to think was no big deal to catch. I felt confident that the wild arowana was finally within reach.

On the last day of August, I flew to the Colombian capital of Bogotá to meet Heiko. It was night when I arrived, and I shivered in the cool air of the Andes as a taxi drove me to a small hotel on a dark street near the city center. Inside, the marble floor shone like glass. I expected to see Heiko sitting in one of the black leather chairs, bleary-eyed beneath the brim of his bush hat. But the lobby was empty except for the woman at reception and the young man she was checking in. The man was very large but gave the impression of a giant baby, with beautiful, dewy skin, a soft, chubby physique, and a sort of disoriented, bewildered stare.

I was just getting my own key when the black phone on the counter rang loudly. The receptionist answered, *"Buenas noches."* She listened for a moment, then, to my surprise, handed me the receiver. It could only be Heiko.

"Hello?"

It wasn't Heiko. It was a man with a Spanish accent who said his name was Humberto. He spoke so rapidly I had a hard time following him, though I caught the gist: Heiko was *not* in Bogotá, as he was meant to be. He was not even in South America. But he sent word that I should proceed to the Amazon without him. He would meet me there.

I had heard that one before.

Up in my hotel room, I opened my e-mail and read, "top urgent message from Heiko." He was stranded in Frankfurt due to an airline strike and said I should fly to the Amazonian port town of Leticia with a Polish man named Michał who was joining the expedition. Come morning, Humberto, the local fish exporter whom I'd spoken to on the phone, would give us a ride to the airport.

That night, I slipped into a fitful sleep, once again dreaming I was on a boat in dark water. At some point, I awoke and realized that the swaying I felt wasn't just a nightmare—it was an earthquake. My bed was actually undulating beneath me. Too exhausted to process yet another crisis, I lay back and let the Andes, which some 11 million years ago rose up and formed the Amazon, rock me back to sleep.

THE POLISH MAN joining the expedition was the giant baby—the one I'd seen checking in. I had kind of suspected as much. In fact, the night before, I'd asked if he was here to meet Heiko Bleher. But he just waved his hand as if rejecting a solicitation. It turned out this was because he didn't speak English. Or Spanish. Just Polish.

Actually, that's not entirely true. He had been teaching himself English all summer in preparation for the trip. Still, whenever I spoke, I felt as though I were physically paining him. After we met in the lobby that morning, we sat side by side in silence, waiting for Humberto to show up to take us to the airport. Finally I asked, "Did you feel the earthquake last night?" I rolled my arms to demonstrate. "Earthquake. Earthquake."

Suddenly his eyes lit up. "Yeah! Yeah!" he said—I had broken through.

Eventually, I got bits and pieces out of Michał. His full name was Michał Babiński (with Michał pronounced as in Mikhail Gorbachev). He was twenty-seven. And he wasn't the experienced jungle trekker I'd been hoping for. Instead, he was a pet-shop clerk from Gdańsk who had never been outside Europe before. In fact, he had never even been camping. He was a fish savant who'd won a trivia contest sponsored by the Polish aquarium manufacturer Aquael, whose name was printed on his black sweatshirt and ball cap. This trip was his prize.

When I asked Michał what he knew about the FARC, he pantomimed chewing his fingers back and forth like the carriage of a typewriter with a terrified look on his face. He said his fiancée, who worked in a butcher shop, was very worried. Then we went back to sitting in silence, both lost in thought—Michał, a large man with a small knapsack, as if he were out for a picnic, and me straddling a backpack that I could barely pick up. I literally had to heave it onto a chair, squat down to secure it to my hips, and then lurch forward into a standing position. I'd had all summer to prepare; and given how many times I'd been warned that Heiko might abandon me in the jungle, I was determined to be as self-reliant as possible. At a sporting goods store in New York, the salesman, a veteran of the Iraq War, had convinced me to invest in something called QuikClot, which he showed me how to pack into gunshot wounds, warning that it sizzled on contact with flesh. I also bought water purification tablets; protein survival bars; a snakebite kit (which consisted of a razor blade, a piece of string, and a note saying to go to the hospital); and a pair of two-inch pods that promised to expand into regular-size towels when moistened. Jeff had rented me a satellite phone, for which I was grateful. I was less grateful for the very large solar panels he insisted I bring to recharge it. I'd tried to conveniently "forget" them under the bed but got caught, so now they were clipped, rather embarrassingly, to the outside of my backpack, where they clanked together as I walked.

"Oh my gosh!" cried Humberto Zea when he tried to lift my bag. A wiry man with a buzz cut, he had rushed into the lobby and announced we were late. Together, we dragged the backpack out to his car, where we leveraged it into the trunk. Michał folded himself into the backseat, and I climbed in up front next to Humberto.

As we sped to the airport, I learned he was a second-generation aquarium-fish exporter who'd known Heiko since the nineties. "He's a very special guy," Humberto said, reassuring me that Heiko had an uncanny knack for dealing with guerrillas. "I don't know how he does it. He knows how to handle the situation." Humberto told me Heiko ate next to nothing on expeditions and had developed an immunity to mosquitoes. "There's millions of them right there where you're going," he added, then shot me a sideways glance. "And they love white skin."

In the past, Humberto had joined Heiko to look for new fish he might introduce to the trade. All the 150 species Humberto currently exported, including the arowana, were caught from the wild rather than farmed. That was the norm for South America. When it came to aquaculture, it was next to impossible to compete with Asia, which produces the vast majority of farmed fish.

Now, however, the annual catch of arowana no longer met Chinese demand. Humberto said the supply had plummeted over the last decade, so almost none of the silvers exported from Colombia actually originated from the country anymore. Instead, they were smuggled out of Brazil, where the species had long been protected by default, since it didn't appear on the short "white list" of animals approved for export.

Surprisingly, Humberto insisted the aquarium trade was not to blame for the sorry state of Colombia's arowana population—rather the US War on Drugs was. Since the 1990s, America had spent billions of dollars not only to aid the Colombian military in fighting the drug-trafficking FARC but also to spray herbicides on coca plantations. Often the crop dusters missed and wiped out large tracts of rain forest. Locals got sick. Though the United States maintained that its chemical cocktail was "practically nontoxic to fish," experts have since contested this claim. The arowana may be particularly vulnerable as a surface-dwelling apex predator consuming other fish directly poisoned by toxins. "The government blames the fish traders,"

Humberto said as he bid us farewell at the airport, "but they're not the problem."

Given all that I'd heard, I began to worry that finding the wild arowana might not be so simple as I'd thought. I was mulling this over on the two-hour flight south when Michał suddenly announced, "*Pseudotropheus socolofi.*" I looked over to see he was pointing at the electric-blue fish on the cover of the book in my lap. I checked the credit. He was right.

The book was the memoir of an American tropical-fish importer who described how in the 1950s the town we were flying to—Leticia—became a hub of the aquarium trade. A major port anchoring the Trapezium, a slab of territory that extends south from the Putumayo and provides Colombia with its sole access to the main course of the Amazon, Leticia can only be reached by boat or plane. Even today, there are no roads in or out.

Upon our arrival, we had instructions to meet someone named Jimmy at the little yellow airport. But no Jimmy appeared to be present. Heiko had given me the name of a hotel that none of the taxidrivers had ever heard of. So Michał and I passed a half hour standing on the curb in the muggy heat of late afternoon. The airport began to empty. The sun was sinking in the sky.

Eventually, we realized Jimmy, a local fisherman, had been there the whole time. He was the skinny man in a soccer shirt sitting across from us on a motorbike. His uncle had sent him to pick up two people who were to accompany Señor Bleher to the wilds of the Putumayo. But Jimmy took one look at Michał and me—the only two people waiting outside the airport—and did not even *consider* that it might be us.

HEIKO FINALLY ARRIVED late that night. I heard him in the hotel courtyard and rushed outside, wondering if I'd hallucinated his voice. But there he was, wearing a khaki shirt, olive cap, and tinted aviator glasses beneath which he looked like hell. Thanks to the strike in Frankfurt, he had spent fifty-two hours in airports and on planes. "These fucking Germans," he said. "They just do what they want." Then he asked: "Where's the Polish guy?"

"Sleeping," I said, expecting Heiko to do the same. Instead, he revealed a new crisis: our flight to the Putumayo had left a day early, and there wouldn't be another for a week. How else we might reach the river was not clear. Heiko wanted to charter a floatplane; but Colombia had all but eliminated these to fight narco-trafficking. He spent the next hour debating options in Spanish with Fausto, Jimmy's uncle, a gray-haired fisherman with a thick walrus mustache whose silver glasses were missing a temple for which he'd substituted string.

Then Heiko announced he was starving, and we walked into town, passing a small park caked in guano where earlier I'd watched a massive flock of screeching parakeets settle in to roost. Eventually, we ended up at a restaurant on the main strip, seated outside amid a cacophony of blaring music and honking motorbikes. Heiko ordered pirarucu, the local name for the arapaima, the monster cousin of the arowana. I'd noticed that the arapaima—not the arowana—seemed to be the star fish around town, featured on wall murals and in the names of businesses. It was easy to see why. Not only was it the largest of the bonytongues but also one of the biggest freshwater fish in the world, roughly the size and shape of a torpedo. You might think keeping such a species as a pet would make no sense at all. However, I had seen it in ponds across Southeast Asia, where it was touted as a novelty, an arowana on steroids. In 2009, when two men mysteriously drowned in Malaysia's Lake Kenyir, the deaths were attributed to the Kenyir Monster, an arapaima rumored to have outgrown an aquarium and been released in the lake. The fish was believed to have rammed the victims' boat.

While this tale was almost certainly apocryphal, the arapaima does have a habit of smashing into predators and prey alike; and at a maximum recorded length of fourteen feet, it can pack quite a punch. These days, however, arapaima rarely turn up more than half that size because they've been severely overfished as the "cod of the Amazon," prized for their boneless steaks. Recently, researchers found that populations were depleted or overexploited at 93 percent of sites examined, and attempts to farm the species remain in their infancy. Still, the fish is ubiquitous on menus across South America, where it's often salted and rolled up like a cigar to preserve it in the tropical heat.

The waiter had assured us that his pirarucu was fresh, but Heiko took one bite and spat it out, declaring it salted. When another plate came, he sent that back too. Disgusted, he announced he would go to bed hungry, and we trudged back to the hotel.

The next morning, we ventured to a military base to inquire about hitching a ride on a cargo plane; but the once-a-month flight to an outpost on the Putumayo had just departed. That left only one option—going by boat, which would require traveling 150 miles down the main course of the Amazon into Brazil, then hanging a left onto the Putumayo, there called the Içá. It was a long haul, but Fausto was encouraging. He said that during rainy season the Colombian fishermen caught the silver arowana in a tributary of the Içá called the Purité, where we could find a giant lake known as El Lago Grande. That lake, he said, contained not only arowana but also discus, the fish Heiko's mother had pursued in the 1950s, and which her son had never stopped seeking in new waters. This seemed to clinch the matter. Heiko announced we would go to the Purité and set about arranging for a boat.

I did not immediately love the idea of crossing into Brazil. I worried that we might not have the proper permits, though Heiko didn't seem concerned. Gradually, however, I decided that the change of plans was potentially a great stroke of luck. By all accounts, the arowana was more plentiful in Brazil, which controls nearly two-thirds of the Amazon. I had to admit that following the advice of the Colombian poachers did seem like the best bet to find the wild fish.

As I was preparing for our imminent departure, however, I heard Heiko cursing in the courtyard: "*Merda!*" I went out and found the boatman he'd hired, a Ticuna Indian named Esteban Meléndez Holanda, explaining in Spanish that it would take another day to get a permit to enter Brazil with as much gasoline as we needed to reach the Purité. An extra day seemed like nothing to me. After all, Theodore Roosevelt spent two months descending the River of Doubt in 1914. Alfred Russel Wallace disappeared into the Amazon for four years in the mid-1800s.

Apparently, neither man had to make sure a pet-shop clerk from Gdańsk was on a flight back to Poland by the end of the week. As part of the deal with the aquarium manufacturer sponsoring Michał's trip, Heiko had purchased his plane tickets and was not keen to pay a change fee, which meant we were crunched for time. We both looked at Michał, who sat playing solitaire on his phone. Heiko rubbed his forehead. "I always have problems," he said. "But this time I *really* have problems."

Among those problems was his failure to acquire a map at the local geographic institute from which he'd returned fuming that morning. Now he asked me to fetch the one I'd brought, studied it gravely, and then plunked his finger on a totally new tributary on the opposite side of the Amazon—the Jandiatuba in Brazil.

All the rivers we'd been talking about lay to the north, but this one flowed into the Amazon from the Brazilian highlands to the south. What appalled me, however, was its apparent proximity to tourist-clogged Leticia.

"It's so close!" I protested, imagining a river crowded with day-trippers.

Heiko laughed. "Follow my finger." He traced the snaking course of the river to its headwaters. "Look how long it is. This is no-man's-land. No one goes there or *has* gone there. There's supposed to be some untouched Indian tribes."

I was skeptical. It seemed preposterous that Heiko, Michał, and I, with less than a week on our hands, could reach uncontacted Indians. There was also no guarantee that this random

new river would have arowana. After all, why would the Colombian fishermen recommend that we travel some 350 miles
to the distant Purité if it were possible to catch the species so
much closer to home? I asked Heiko these questions, but he
waved away my concerns. "The arowana is in the entire Amazon basin except for above waterfalls," he said. "Don't worry.
We'll catch lots."

THAT NIGHT, I awoke to torrential rain hammering the
roof. It was still pouring at 4:00 a.m. when I met Heiko and
Michał in the flooded courtyard, and we loaded our supplies
into the back of a taxi. We drove through the dark town to
the port, where we tottered down slippery planks, bouncing
precariously, as we carried everything onto the boat. Most of
Heiko's equipment fit in an aluminum trunk so dented, duct
taped, and papered with stickers that I imagined it had been
traveling with him for decades.

I was wrong. It was a year old.

One thing Heiko did not bring was bottled water, as he said
he always drank straight from the river; and he grumbled about
how many gallons I'd convinced Michał we needed to haul
with us for proper hydration. Even more dismaying to Heiko
was the size of the boat, which wasn't the one he'd selected.
He lamented that it was too big to navigate small creeks. But it
was too late to change it. "We will just have to borrow a canoe
from the Indians," he sighed.

To me, the wooden boat didn't look so big. Painted bright
blue with yellow trim, it had a narrow hull lined with two
benches on which we now sat under the cover of a tarp. Perched
at the stern, the boatman Esteban had no such shield from the
elements but gave us a warm smile before revving up the out-
board motor. A whiff of diesel fumes sputtered into the air and
we were off, headed down the Amazon in the dark, the cold

rain stinging our faces behind thick canvas ponchos. Heiko's was royal blue like a wizard's cloak, with a pointy hood lined in the same silver as his beard. Seated near the helm, he leaned forward, his arm resting on one knee, and stared out beyond the bobbing prow like Captain Ahab pursuing the white whale.

When it comes to iconic quests for fish, the English literary canon is decidedly skewed toward salt water. The most famous narrative—*Moby-Dick*—is not even about a fish at all, but rather a mammal. Heiko resented that freshwater species were so sorely underrepresented not only in the popular imagination but also within the world of science. "Ichthyologists that specialize in freshwater, there are very few on this planet," he told me. Rivers and lakes might be the hotbeds of evolution—the cradle of vertebrate biodiversity—but the mighty oceans were sexy. In the 1980s, Heiko had met Jacques Cousteau in the Brazilian city of Manaus, where the French oceanographer was filming a documentary series on the Amazon, one of the few times he strayed from the sea. "Very snobby," Heiko recalled, grumbling that Cousteau's film did not feature a single discus. "He never got it about freshwater."

More recently, the BBC had invited Heiko to London to consult on material for the series *Planet Earth.* "I told them about the most fascinating life-forms on this planet," he said, giving the example of a fish called the golomyankas, native to Siberia's Lake Baikal, the oldest and deepest lake in the world. He explained that the species lives at a depth of more than three thousand feet, where the pressure makes it impossible to give birth. Consequently, the pregnant female swims toward the surface, where she promptly explodes, releasing her young to return to the bottom. In the end, Heiko said, the producers didn't think the public would be interested in such small fish. They focused on dolphins instead.

As dawn broke, the storm passed and the slate-gray sky began to clear. Heiko poured hot coffee from a thermos and handed out banana sandwiches. We rolled up the rain tarp, exposing the boat's rib cage, and peered out at the wide, brown river, expansive mudflats extending from either bank. Despite the morning's downpour, it was dry season and the water was unusually low—drier than Heiko had ever before seen the Amazon. Because of this, there were no large riverboats. When a crowded speedboat zipped by, Esteban called out in Spanish that it was the bus to São Paulo de Olivença, the small town where we planned to hang a right onto the Jandiatuba.

"It always amazes me when people say they're taking an Amazon cruise," Heiko said, pointing out how deforested and eroded the main trunk of the river is. Its ruination, he told me, was basically the work of one generation. Over the past fifty years, more of the Amazon rain forest has been clear-cut and burned than in the previous five centuries. Ranchers, farmers, and timber barons running industrial-scale operations have executed much of the destruction. More recently, small homesteaders have been encroaching upon the forest.

With each of his expeditions, Heiko told me, he had to travel farther and farther, to ever-more-remote corners to find untouched nature. By last official count, the Amazon has more than a thousand tributaries. In his search for wilderness, Heiko claimed to have explored practically every major one. "This one we're going to now, it is the only one I have not been deep into," he said.

With that, he stretched out in the warm sun for a nap, his arms folded under his head, the brim of his cap pulled down over his sunglasses. Michał followed suit with headphones on. I sat across from them and thought about Heiko's claim that he was, in essence, about to *finish off the Amazon,* or at least its main tributaries. Doubtful as this seemed, the idea made me melancholy, as if the world were an aquarium, and we had reached the glass wall.

Heiko once surprised me by saying he felt as if he lived in

the best possible time to be an explorer. Modern transport allowed him to reach places in a few days that would have taken his predecessors months or years to get to. "I can do sixteen expeditions in a year, which they couldn't do in a lifetime—not even my mother," he said.

"But isn't it a little sad that you can get to these places so easily?" I asked.

"Well"—he mulled this over—"it would have been adventurous, naturally, to travel like Wallace. But now you can see a lot more, you can understand a lot more."

As the day wore on, and Heiko arose from his nap, the heat of the afternoon eventually began to fade. The later it got, however, the stranger it seemed that we had not yet reached the Jandiatuba, which had looked so close on the map. At five thirty, the sun was setting behind us; by six thirty it was completely dark. Heiko climbed up to the prow, where he scanned the water with a rickety spotlight that flickered on and off, so Esteban could see where he was going from the stern. Ours was the only boat on the river now. We'd been traveling for more than twelve hours. Sheet lightning crackled in the distance.

Just after eight, we finally spotted a faint fluorescent light near the shore. As we approached, I could make out a floating dock: planks glistening with algae, a man sleeping in a string hammock, and a sign painted with the name of an unfamiliar town in red script.

Heiko saw that sign and roared, *"Esto no es São Paulo de Olivença!"* While he'd been napping, Esteban had missed the turn. We'd overshot the Jandiatuba *by more than seven hours.* We had reached the Içá—the Putumayo—after all.

I WAS THRILLED. Heiko was not. I felt as though fate had delivered me to the arowana's doorstep. Heiko thought we were in a real mess. We had reached the town of Santo Antônio do Içá, which sits at the mouth of the Içá River, and which Heiko declared the "worst place on the Amazon." He said he had once counted five rats in his hotel room. He said he would prefer to be in jail.

Continuing up the Içá to the Purité would require at least another twelve hours of travel. Heiko wasn't sure we could even get enough gasoline to make it there and back. He asked to see the map again and kept looking from it to me, and from me to it. He sighed. He traced his finger one way and then the next.

That night we slept stretched out on the benches of the boat, mosquitoes whining in our ears. It was still dark and Michał was snoring when I awoke to the sound of Heiko's clapping for the attendant in the hammock to sell us gas. I looked at my watch: 3:57 a.m. Sitting up, I heard Heiko instruct Esteban to continue on to the distant Purité, where the Colombian fishermen had said we would find the arowana, and forget about returning to the much closer Jandiatuba, with its uncontacted Indians.

"I just did not want to go back," he said when he turned to see me watching. Maybe that was true. Or maybe he did it to be nice because he knew how desperately I wanted to find the arowana—and the Purité seemed like our best shot.

Either way, we now had a long journey ahead of us. As the sun rose, freshwater dolphins as pink as Pepto-Bismol surfaced here and there. I could see the Içá was far more verdant than the Amazon itself, though it was still a vast river. "It's one of the smaller affluents of the Amazon, and it's so *big*," Heiko said appreciatively. He had last been here in 1977, when he'd made one of his proudest discoveries—the first red discus, a brilliant fish with white markings and blue fins that adorned the black T-shirt he now wore. He told me he'd found a single specimen, which he brought back and bred with other varieties, the fish becoming the progenitor of red discus everywhere, as it was never again found in the wild.

From time to time, we passed clusters of wooden houses built by poor farmers from the coast who came to squat on the land. Heiko stopped at these tiny settlements, climbing up steep inclines to ask in Portuguese if anyone had seen the fish on his shirt. One man standing in a doorway surrounded by five small children assured Heiko the discus could be found in the Purité. His mood brightening, Heiko thanked the farmer and handed out orange candies to the children, the smallest of whom sucked on the plastic wrapper as an insect buzzed around her fingers.

Back on the boat, Heiko carefully recorded the name of the settlement on my map, which he set down on the seat. Suddenly, a gust of wind blew it into the river. "Shit!" he cried, clambering over the bags to get to the stern. Esteban stopped the motor and swung the boat around. But it was no use—our only map had sunk.

I could have cried. Given our track record *with* a map—how lost we'd gotten the day before—I doubted we would ever reach the Purité now.

CHAPTER SEVENTEEN

Here Be Dragons

BRAZIL

But we did reach the Purité. It was late afternoon when we at last arrived at the mouth of the narrow river, walled in by a thick snarl of vegetation. There were no signs of human life, though there were plenty of fish, which leaped in the air at the sound of our boat. "To announce they're happy we're coming," Heiko said. "I believe this—that fish jump for happiness. Like dolphins. Like humans."

Esteban steered into a small lake off the main artery and docked at a gap in the foliage. When he turned off the motor, a perfect silence fell over the place. Heiko broke it with a splash, diving into the black water, having stripped down to a pair of blue underwear. With a theatrical air, he breaststroked across the lake, then disappeared under the surface, popping back up, his head glistening like an olive. "No one wants to bathe?" he called back to Michał and me. "The water is fantastic!" After a few more laps, he trudged out of the muck and shook off like a wet dog. "So you see I got in naked where there are thousands of piranhas, and they don't eat me."

Earlier that year, Heiko had written the "Definitive Guide to Piranhas" for *Practical Fishkeeping*, which began, "It's a complete myth that piranhas are man-eaters." Heiko did not believe that *any* animals ate people, unless perhaps they were already dead. When a crocodile was found to be digesting a man in Australia, Heiko said the guy was clearly murdered and then fed to the crocodile to cover up the crime.

In the case of piranhas, Heiko was not alone in coming to the fish's defense. The piranha's bad reputation can be traced to Theodore Roosevelt, who called it "the most ferocious fish in the world," following the 1914 Amazon expedition that nearly killed him. More recently, however, numerous ichthyologists have argued that the fierce reputation of piranhas is overblown, pointing out they mostly eat the fins of other fish and are actually rather timid.

Tyson, for one, disagrees. "Ichthyologists are in a state of denial," he told me. "They'll admit that yes, if you dangle your hand in the water, a piranha might bite your thumb or your little finger off. But they won't admit that if you walk in the wrong place, and there are lots of hungry piranhas around, there's not going to be anything left of you." He showed me graphic photographs of a man allegedly killed by piranhas in Brazil. "Now"—he pointed helpfully—"they removed the scalp, removed the lips, removed the eyeballs." According to Tyson, piranha apologists claimed the fish would never eat the eyeballs. "That's nonsense," he said.

One fact, however, was not up for debate: the Purité was *teeming* with piranhas. When Esteban set up a fishing line over the side of the boat to catch some dinner, piranhas were all he got. They were about the size of my hand, silver with flaming-red bellies, and had a fearsome set of serrated teeth, which Heiko showed me by using a stick to pull down the bottom lip of one. He plopped the fish in a small glass display tank, then instructed Michał to stick his hand in and hold the wriggly piranha still against the side of the glass so he could get a good photograph. Michał looked terrified, but Heiko reassured him, "Just don't put your hand near his mouth." He then pointed to a large white scar under his thumb. "One like this once bit a chunk out of my hand. He was just trying to defend himself."

The piranha is not the only fish to worry about in the Amazon. There is also the notorious candiru, a catfish the shape of a toothpick, the only vertebrate other than the vampire bat known to survive solely on a diet of blood. Ordinarily, candiru swim into the gill chambers of larger fish, anchor themselves with tiny spines, suck a blood meal, then swim back out and burrow in the riverbed to digest. However, the candiru sometimes gets confused, perhaps mistaking a stream of urine for the gill stream of a fish, and enters the wrong orifice—a penis or a vagina or even an anus. This is bad for both the candiru, which is fated to die, and for the human, who now has a dead candiru lodged in a most inconvenient place. In 1897, the ichthyologist George Boulenger presented a specimen of the candiru to the Zoological Society of London and announced, "The only means of preventing it from reaching the bladder, where it causes inflammation and ultimately death, is to instantly amputate the penis."

Much has been made of such penectomies, though the veracity of these cases remains sketchy. Heiko didn't believe any of it. He suspected Indians came up with the story to keep white men from peeing in their drinking water. Later, when he found a candiru attached to the gills of another fish, he showed me its tiny spines before dropping it back into the water in which he stood waist deep. That, only a few minutes after an electric eel popped up next to him like a periscope, taking a gulp of air, which is how the fish breathes. How the fish eats is by shocking its prey with up to 860 volts—technically enough to kill a person.

But Heiko was adamant that being in the jungle was safer than crossing the street in New York. He said all the horror stories about dangerous wild animals were a Hollywood fabrication from the old Tarzan films of the 1930s and '40s. "You don't have to be afraid of anything," he told me, as he waded through the piranha-filled water with his hand net, straining the muck for small fishes. He examined them one by one, tossing most over his shoulder, shaking the suckerfish free from his fingers. Anything he found of interest he put in a water-filled plastic bag, where it would remain until he photographed it, then released or preserved it. "I photograph everything to catalog what is here," he explained.

While Heiko collected, Esteban started a fire to boil water for dinner and then helped me string my jungle hammock between two saplings. It was a contraption with a mosquito net that peeled open like a banana, which I'd bought in New York and practiced hanging in Washington Square Park until a helpful drifter warned I was bound to get in trouble with the law. Despite recommending a hammock, Heiko had brought tents for himself and Michał, which they set up nearby. By seven, it was totally black, the darkness punctured only by stars overhead. We sat on the leaf litter, eating boiled piranhas, Heiko's favorite fish, and listening to frogs that sounded as if they were hopping around on pogo sticks.

Until recently, our campsite had been underwater, this being one of the Amazon's great flooded forests, inundated for about six months each year. During the rainy season, both the arowana and discus rear their young amid the trees. Now that it was dry season and the waters had receded, the fish would theoretically be easier to find.

After dinner, we took the boat out onto the dark lake to begin our search. Heiko leaned over the bow, toes curled, inspecting the water with a flashlight, netting something interesting here and there. Finally he said to Esteban, *"No hay nada."* There were no discus here. No arowana. To find those, it seemed we would have to reach the big lake that the old fisherman Fausto had told us about.

EARLY THE NEXT morning, we set off upriver. From time to time we stopped at small streams, which the boat was too big to enter. While Michał and Esteban stayed on board, Heiko strode through the forest, shoeless, with me tripping along behind him. "It is such a fantastic feeling to walk barefoot," he said during one such foray, stepping over an armadillo hole. "You feel the nature. You feel where you are."

He climbed down to a creek and began sorting fish in his hand net. "An *Apistogramma* under every leaf," he breathed, referring to a popular pet the size of a paper clip. He had already discovered what he thought was a new species of the fish the night before—a tiny creature with blue spots, red-tipped fins, and a black line bisecting its body. Like Heiko's beloved discus, *Apistogramma* are members of the rapidly evolving family Cichlidae. In recent years, a divide has emerged among conservation biologists regarding which species are most important to protect—ones from such active lineages, whose preservation secures the future of adaptation and speciation, or the oldest species, such as the arowana. On the one hand, the arowana is unique to the world, the remnant of a bygone era. On the other, it's an evolutionary dead end, having changed little in more than 100 million years. Which is more important—holding on to life's history or facilitating its future? To me, it seemed an impossible choice.

Because it was dry season, all the little creeks were shriveling up. In one that was little more than a trickle, Heiko found about 150 fish in a single scoop of his net. "Can you imagine such little water and so many fish?" he said, explaining that in weeks the creek would dry up, and all these fish would die, having deposited their eggs in the mud, where they would remain until hatching with the coming rains.

Back in the boat we continued up the Purité, here and there passing white sandy beaches—vestiges of a time when the entire western Amazon was a vast inland sea. Midmorning, Heiko selected an especially pretty spot at which to stop and photograph the fish he had collected, dragging his aluminum trunk onto the sand and creating a backdrop of green branches into which he nestled a small aquarium. When he asked me to bring him a gallon of our limited drinking water, I happily obliged, thinking the scorching heat had finally gotten to him. Then I heard *glug, glug, glug,* and turned in horror to see him filling the aquarium, declaring the black water of the river too dark. He knelt down in the sun, his black cap turned backward, crooning to the fish like a high-fashion photographer, "Spread those fins" and "Come on, sweetheart, I know you love me" and "Yes! Yes! Yes! Yes!"

Esteban cast a line from the boat and before too long called out excitedly in Spanish. He had caught a stingray the size of a sombrero, brown with black-ringed spots and a spiked club of a stinger. He tugged the fish onto the beach and lifted it up by the gills, located like a handle above its strangely sensitive eyes. The Amazon's stingrays, like its dolphins and white sand beaches, are yet another relic of its distant past connected to the Pacific. Recently, several stingray species had become popular in the aquarium trade, fetching prices rivaling those of arowana.

The reason Heiko wanted to catch a stingray, however, was that Tyson had requested specimens. "Tyson is going to be very, very happy," Heiko said, inspecting the stingray on the beach. "I have never seen this one before." He asked me to take a photo of him with the fish, which promptly stung him in the hand (though Heiko, wincing in pain, said it was only defending itself). For the sake of science—not revenge—he announced that he had to kill the poor creature. He maneuvered it into a large plastic bag and poured in formalin, the remainder of which he then stowed alongside the bottles of water. "Don't drink this," he said, "or you're going to be dead forever."

Just then, we were startled by the unexpected sight of a large red-and-white riverboat sailing around the bend, heading upstream. It was crowded with people dangling over the sides of its upper and lower decks, some of them waving as they passed. We hadn't seen *anyone* on the Purité, which seemed to be entirely uninhabited. Who were these people? Where were they going?

The answers came later that afternoon when we heard a strange chanting piped through loudspeakers and glimpsed a settlement built high on a clearing above the river. A woman doing laundry on the muddy bank explained that it was a religious community. Until recently, fifty people had lived here. Now a hundred more had just arrived. Heiko had seen such cults hidden all over the Amazon. Disgusted by the presence of settlers, he refused to go up and say hello.

By now, it was late afternoon, the time of day when Heiko expected to see parrots flocking toward the sunset, but we glimpsed only a few scattered pairs. "It's so disappointing that you don't see any birds," he said. Heiko was convinced that the Purité had been heavily hunted—which made sense, he reasoned, considering we had come here on the advice of Colombian poachers. "From the point of view of animals, I'm very disappointed. People must have done a lot here. Otherwise there would be more."

He began to think of his two-year-old daughter back home in Milan. "I want to take Amanda to see the jungle while it still remains. The next generation won't see anything."

That night, Heiko led Michał and me though the forest to look at insects, but we didn't find much. When I drew his attention to a curious brown bug with a black head, he told me it was just one of the Amazon's wild cockroaches. The next day,

Heiko spotted a puma print in the mud. A while later, when we heard some rustling, what came rushing out of the brush was not at all what we expected. "In the deepest jungle, we find a wild animal—a dog!" cried Heiko.

The dog was yellow, so skinny she looked like a walking rib cage, but her tail wagged nonetheless. We were no longer close to the religious community, so Heiko thought the dog must have been abandoned by hunters. "Maybe there is some old bread to give her," he said, but I couldn't find any. When he turned back to his fish, I dug out my secret stash of protein survival bars and fed them to her one at a time. Esteban leaned over the side of the boat. *"Cómo se llama?"* he asked the dog. *What's your name?*

According to the zoologist James Serpell of the University of Pennsylvania, early humans may well have loved tamed wolves as pets before using them for any other purpose such as hunting. He argues that animal domestication—which easily rivals the invention of stone tools, religion, language, mathematics, and even modern technology for its influence on human culture—could have begun with our simple desire for the companionship of animals. Even today, many tribal societies bring home the young of other species to suckle and rear.

A competing theory has recently gained traction among scientists that dogs, like cats, may have domesticated themselves. Maybe wolves with a certain temperament began to hang out near human camps to forage for scraps and then bred with other animals that did the same, self-selecting for tameness over time. The remarkable truth is that we, as a species, appear to have done something similar. In evolving from our apelike ancestors, modern humans have developed the same suite of physical traits that domesticated animals invariably display: shorter faces, smaller teeth, more feminine features. It seems, in some mysterious way, we have managed to self-domesticate. Perhaps in fish we recognize the wildness we lost—even as we're now destroying it.

Late that afternoon, I was sitting next to Heiko at the helm when the motor coughed and died, a tranquil silence falling over the river. As we waited for Esteban to refill the gas tank, Heiko crossed his arms and looked up at the sky, which was now a dusky shade of periwinkle. He told me that when he was a child in the Amazon, motors didn't exist. "Only paddles," he said.

The nostalgia in his voice touched me. Suddenly, I felt terribly sad. When I first set out to write about the arowana, I had been attracted to the humor and high drama of the fish world, to the eccentricities and obsessions of the people who were part of it. But there was no way to think about the arowana—about any fish, really—without confronting loss on a scale that seemed too large for the human mind to comprehend. I had come so far to find one wild thing, to experience the wild itself, and all I had to show for my quest was a cult, a cockroach, and a starving dog. Despite myself, tears welled up in my eyes and spilled down my cheeks.

If Heiko noticed, he didn't say anything. He kept looking at the sky. "So peaceful," he said. "You're probably going to miss this when you get back to New York."

DARKNESS WAS FALLING when we reached it—El Lago Grande. The Big Lake. In the dusk, I could just make out the tangled green shores, dark and inscrutable, casting shadows that bled into the water like dye. This was the lake from which Fausto and Jimmy had told us Colombian fishermen caught tens of thousands of arowana, smuggling the young out of Brazil each year during rainy season. "It looks like the arowana are coming to the surface," Heiko said, as he stood at the helm, surveying the indigo waters. I strained my eyes to see what he saw, but it was futile in the fading light.

We set up camp above a muddy incline, while Heiko stayed

on the boat to photograph fishes he had caught earlier that day, the beam of his flashlight illuminating the aquarium like a golden reliquary. One of these fish Esteban boiled for dinner. "You can say you ate the beautiful *Geophagus camopiensis*!" Heiko said to Michał.

"I don't have problem with that," Michał said.

Around nine o'clock, we climbed aboard the boat and set off in the velvet night, my heart thumping in anticipation. Since the motor would scare away the fish, Heiko sat in the front to row while Esteban took up the rear. After ten minutes, however, we had barely budged, and Heiko fumed about the size of the boat. Another ten minutes passed, and he was still smiting the lake with his oar, each stroke inching us forward almost impercepti-bly. It took half an hour to reach the nearest mangrove, which we promptly crashed into. "Never in my life have I collected discus in such a big, clumsy boat!" Heiko said. "Practically impossible."

Nevertheless he lowered himself onto his stomach and leaned over the bow, shining his flashlight into the cloudy water. I could see the bottoms of his feet, which were stuck full of palm spines he'd accidentally trod upon during his barefoot forays on land. He remarked in Spanish that the water didn't look deep, which was not promising since discus like to school at great depths. "I doubt we'll find discus here," he said. "I really doubt it. I don't think there *are* discus here."

We had no choice but to turn on the motor and search for a less shallow part of the lake. As we sped across the water, Heiko shone the spotlight on South American alligators called caimans. "Snake," he said, illuminating a pair of red eyes dangling from a tree. A big animal toddled through the brush, and Heiko guessed it was a capybara, the world's largest rodent, or a tapir, which looks like a pig but is more closely related to a horse or rhinoceros. Michał searched for these animals too until he shined his flashlight one too many times in Heiko's eyes, which put an end to that.

Eventually, we reached a second mangrove and Heiko disappeared over the prow again. "Ahhhh," came his voice from below. "Stop!" He had found angelfish, the original "king of aquarium fishes" from his grandfather's day, before the discus knocked it off the pedestal that would in turn be occupied by the arowana. He netted one of the silver triangular creatures with four vertical black stripes and handed the net to Michał, who tried to extract the fish with his hand.

"No, you don't do it that way," Heiko said impatiently, taking the net and flipping it over so the fish plopped easily into a cooking pot filled with water, which he was using as a makeshift holding tank. "You should always in your life avoid collecting fish in your hand—they have a very sensitive slime on them."

For a long time, he continued peering over the edge, muttering in Spanish and scooping up small fishes. Finally he blew out a puff of air with disgust and said halfheartedly, "Do you want to catch a croc?"

Esteban paddled us toward a young caiman, but the boat proved too unwieldy. We plowed into the mangrove, and the caiman got away.

I thought that *now* we would finally look for arowana. Instead, Heiko abruptly called it a night. "*Vamos,*" he said and pointed Esteban back to camp. Panicking, I asked about finding the arowana. With an air of weariness, Heiko told me he'd been keeping an eye out but hadn't seen any.

Night was the time to catch the fish, when it floated near the surface of the water, its guard down, its eyes reflecting the glow of a flashlight. Because we had come so far—and Heiko had to get Michał back for his flight—we had just this one night on El Lago Grande. It seemed to me Heiko had given up far too easily. What about all this talk of his staying out past midnight, diving in black water? It felt as if he were punishing me because the lake did not have *his* fish—the discus. "Do you think it would be possible to see the arowana during the day?" I asked after we returned to camp.

"The problem is this lake is gigantic—if you noticed, huh?" Heiko said, brushing his teeth over the side of the boat, gargling the black water. "And arowana are solitary fishes, no?"

Overhearing our conversation, Esteban looked up, his face illuminated by the lantern he was holding. *"Quería arowana?"* he asked, nodding in my direction. *She wanted arowana?*

"Sí," Heiko said. *"Difícil, no?"*

Esteban nodded silently—very difficult.

I BARELY SLEPT, listening to the mad improvisation of insects. I could feel the hours of darkness slipping away, along with my last chance of ever finding the wild arowana. At dawn, I plopped out of my hammock, mucked through the soup of rotting leaves, and watched the golden orb of the sun rise over El Lago Grande, feeling dejected and dispirited. The sky was clear and the coolness of nighttime lingered in the air. But the tranquillity of the scene just reminded me that the night was over—the jungle had gone to bed. My best opportunity to find the arowana was gone.

Esteban was next to stir. He climbed over the edge of the boat, where he had slept, and flashed me a bright smile as he trudged through the muck to collect kindling for a fire. Then Heiko emerged from his tent, grunting in my direction. Over coffee, he announced that he still had fish to photograph, saved in pots from the night before. In the meantime, he asked Esteban to take me to see if we could find an arowana. Faced with the prospect of posing more piranhas, Michał opted to come too.

After breakfast, the three of us climbed into the boat, leaving Heiko behind, and set off across the large lake to the opposite shore of mangroves, where Esteban cut the motor and let us drift for a while. It was a beautiful day. The dark water reflected the blue sky. Two white birds fluttered inside the gnarled mass of trees and something lizardlike skittered across the water.

After a few minutes, Esteban held up one finger to his lips, stepped gingerly to the bow, and lifted the green cast net lined with orange bobbles. He stood frozen for a long while, his eyes fixed on the water. Then in one sudden, smooth motion, he hurled the net, which landed with a loud splash, flushing the squawking birds from the mangrove.

Something about Esteban inspired confidence. I believed in him, this kind, handsome, good-humored Ticuna Indian, who slowed the boat whenever he saw me trying to put in my contacts and losing them to the wind.

Now he drew in the net and lifted it up, dripping. It was empty. He frowned and showed me the problem. In addition to its being daytime, totally the wrong hour to catch an arowana, and our having a clunker of a motorboat, when everyone knows you're supposed to sneak up silently on the fish, our net was chomped full of holes. "*Muchas piranhas*," Esteban said, shaking his head ruefully.

Nevertheless, he tried again, this time catching a small silver fish missing its tail. Next came a large, gaudy catfish decked out in leopard print like an eighties aerobics instructor.

As I stood next to Esteban, watching his deft handling of the net, calmed, despite myself, by the warm sunshine on my head, I considered the situation. Once upon a time I had wanted to find out why a pet fish was so irresistible that people smuggled it into the United States, risking their very liberty. Three and a half years and fifteen countries later, I was now in Brazil (possibly illegally) pursuing the fish myself. At some point, things had gotten out of hand.

But this was supposed to be the most reasonable leg of my quest. Unlike in my previously overambitious attempts, this time I had set my sights on an arowana that was *not* officially endangered and lived in a lake where poachers caught the fish regularly. Moreover, I wasn't on my own, but rather in the company of Heiko Bleher, the world-class explorer who claimed to have single-handedly started the arowana trade in the 1960s.

Yet I *still* couldn't find the fish.

I thought about Heiko's mother, Amanda Bleher, who never found the discus that she dragged her children through the green hell to acquire. I thought about Alfred Russel Wallace standing on a lifeboat in the middle of the Atlantic Ocean, watching four years' worth of collecting burn up in the fire that destroyed the ship carrying him back to England. In his arms, Wallace clutched the single tin box he was able to rescue from the flames. As it happened, it contained his drawings of Amazonian fishes, among them a delicately rendered sketch of the arowana. Half a century later, he was still trying in vain to publish these drawings, because no museum wanted them without the lost specimens.

On the face of it, the Amazonian expeditions of Amanda Bleher and Alfred Russel Wallace were both spectacular failures. Yet I doubt that either of them would have opted out of their grand adventures, given the choice. It was the same for me. So what if I'd failed? It was never about finding the wild fish—not really—but about adventure and exploration and understanding, all of which I had gained in abundance.

Esteban gave me a weak smile. Our efforts were obviously hopeless. I nodded my agreement. It was time to quit and head back to camp, to begin the long journey home. I sat down next to Michał, and Esteban climbed to the stern. He flipped on the motor, which roared to life, cutting through the quiet. At the startling sound, something shot out of the water like an exploding champagne cork. "Arowana!" Esteban cried, wagging his finger at the water.

I take it back! It was always about the fish!

The three of us leaped up and stood there in disbelief, staring at the expanding rings of water where the silver streak of a creature had just plopped. Then Esteban got an idea. We would not tiptoe around and sneak up on the arowana, as traditional wisdom dictated. No, we would do the opposite. He flipped on the motor again, steered the boat right at the mangroves, and began racing along their edge. Michał and I beat our hands on the metal gas drum, hooting and hollering. Suddenly another arowana leaped alongside the boat. Then another, a monster of a fish, traced a double arch through the air. Maybe three feet long, it glistened silver with a rosy-hued belly. When it sprang into the air, it left the water entirely and was, for a moment, suspended at eye level, looking at us as we looked at it.

"Wow!" said Michał, who was not prone to such effusive displays.

Esteban beamed.

And me? I was reminded of a story my friend Amos Yu in Borneo told me. Amos loves plants—he runs a nursery—and he'd become obsessed with finding one species in particular: *Nepenthes platychila*, a red pitcher plant that traps and digests insects in special jug-shaped leaves. The plant was thought to be endemic to the virgin rain forest of Borneo's remote Hose Mountains. So Amos made a costly and arduous expedition there. He found a begonia that hadn't been found in a hundred years. But no pitcher plant. He went back a second time and scaled two waterfalls to reach a sheer sandstone wall where the pitcher plant was supposed to grow. No pitcher plant. He went back a third time and made it up to a third waterfall. Still, no pitcher plant. On his fourth trip, after spending eight years and $20,000 in pursuit of this plant, he found it on the very first day, growing in a tree. It wasn't especially beautiful, and he would never even try to sell it. Yet he was so overcome with emotion that he almost cried. He felt joy and elation and exaltation—profound gratitude for being born into such a strange and wondrous world—but more than anything else, immense relief. In finding the thing he'd sought for so many years, he was liberated from its tyranny over his life, released from the spell that had bound him. As he admired the modest little plant, which he realized was just that—a plant—his first thought was "I don't need to come back!"

I felt exactly like that. On the long journey home, we would run out of drinking water, which Heiko had been using to fill his aquarium, and then out of gas, which some friendly Ticuna Indians replenished, refusing payment. When *that* gas ran out, putting Michał at risk of missing his flight, Heiko ran to the bow, grabbed an oar, and began paddling madly up the Amazon.

"Oh, Jesus," said Michał, looking up from playing Sudoku, then returning to his game.

"You know," Heiko called out to me, between heavy breaths, as I searched in vain for another oar, "this always, always, *always* happens. I never had any of these nine hundred expeditions go smoothly."

Through it all, I took solace in the knowledge that I had found the object of my quest, a fish that turned out to be no more or less marvelous than the great diversity of life surrounding it. The arowana was, despite everything, just a fish, as ordinary and extraordinary as that may be.

For now, at least, it was still there—in the wild.

Alfred Russel Wallace's sketch of the silver arowana, 1851.

Epilogue

For my sixteenth birthday, a friend gave me five goldfish in a crowded bowl. Four died in rapid succession. The fifth, which I named Stephen, lived—and lived and lived. Sprawled on my bed, I'd contemplate his little egg yolk of a body atop my bookshelf. Now that he was alone, he didn't swim around much but sat on the glass floor, apparently depressed. I saw a parallel between his life and mine: me in my room, him in his bowl, neither of us exploring the wider world. The only difference was that *he* didn't have a choice.

One day, when it became clear Stephen was around for the long haul, I decided to do something to spice up his sad existence. I drove to a pet shop where the clerk pointed me to an aisle filled with plastic castles and skeletons blowing bubbles. I didn't think any of these artificial trinkets would cheer up Stephen. Instead, I told the man I wanted to buy a real aquatic plant—the kind a fish might encounter in nature. Shrugging, he disappeared into the back and reemerged with a dripping bag of long green fronds.

I don't know how I expected Stephen to react when I installed this greenery in his bowl. But his evident delight could hardly have been more satisfying. He swam in circles of what I took to be rapturous ecstasy. When he wasn't twirling around the leaves, he rose to the surface to give my pinkie tiny kisses of gratitude.

The next day, he was dead.

I later learned that the plant probably introduced some lethal bacteria. Stephen hadn't been pirouetting in joy but rather writhing in agony. His butterfly kisses had been gasps for breath.

"The plant killed my fish," I told the Finnish aquarist Tor Kreutzman, recounting this story fifteen years later.

"No, *you* killed your fish," he said.

He had a point. Dominion over other creatures is a fraught responsibility. Even the best of intentions can go horribly wrong.

AFTER I GOT back from the Amazon, I learned Heiko was right: there really *are* uncontacted Indians living at the headwaters of the Jandiatuba, the river we narrowly missed while he was napping. They're called *flecheiros*, or the Arrow People, because pretty much all anyone knows about them is that they shoot poison-tipped arrows at outsiders who dare trespass on their territory.

It was always like this with Heiko: just when I was sure he was full of it, his hyperbolic claims would turn out to contain at least a kernel of truth. For example, in the foreword to his mother's memoir, he writes that she was "possibly the most extraordinary woman of the twentieth century"—not only an incomparable naturalist but also a tennis, Ping-Pong, ice-skating, and roller-skating champion, as well as the winner of more than 148 motocross races ("the only woman competitor"), and one of the first women to venture onto a motorbike, fly a glider, and drive a car.

This seemed a little over the top.

Yet when I visited Heiko's nineteenth-century villa outside Milan, he hauled out a dented aluminum suitcase and an army-green duffel bag containing Amanda's papers, mildewed from her years in the Brazilian jungle. Inside the mass of yellowed pages, I found photographs of a young woman twirling through the air in roller skates, racing motorbikes with goggles atop her head, and canoodling with an ex-paramour identified as Bob, who Heiko mentioned was Hitler's secret right-hand man.

Amanda kept extensive journals, and many pages were scrawled on a crumbling, splinter-strewn substrate that Heiko speculated was toilet paper from her three years behind bars in a Frankfurt prison on murky espionage charges. I stayed up the entire night photographing whatever I could. Later, when I enlisted a German scholar to read through this material, he confirmed that Amanda did appear to be writing on toilet paper from the post–World War II era. He said it was not entirely implausible that Hitler could've had a right-hand man named Bob.

So I've learned to give careful consideration to what Heiko says, even when it seems far-fetched. As for his conviction that protecting animals leads to their inevitable demise, however, I can't agree. Consider the crocodiles brought back from the brink of extinction by CITES. Or the Devils Hole pupfish, the first fish listed on the US Endangered Species Act in 1967, which lives in a spring basin the size of a walk-in closet in Nevada. Without extensive human interventions—a federal hatchery, artificial refuges, and battles over water rights going up to the Supreme Court—the species would surely be gone. Instead a population of a few dozen persists on life support.

At the same time, Heiko's claim contains a kernel of truth: declaring an animal endangered *can* make it more desirable and thereby increase its exploitation, which is likely what happened to the Asian arowana. Yet the more problematic issue may be how we determine which species to protect in the first place. How do we prioritize one animal over another when so many are in dire straits? The warm and cuddlies generally garner far more attention than less charismatic creatures like fish. But the lesson of modern conservation is clear: the only approach that works is holistic—preserving ecosystems in their entirety.

Interestingly, Brazil has reached a similar conclusion about the Arrow People. While the government once strove to protect isolated Indians by establishing contact and regulating their existence, this policy inevitably spelled the destruction of their way of life—not least by introducing deadly germs to which they had no immunity. In the late 1980s, officials reversed course and adopted a policy of *not* making contact with such tribes, instead roping off the forest that sustains them and preserving the system as a whole. It turns out the river on which the Arrow People live—the Jandiatuba—flows through the Javari Valley, a protected reservation twice the size of Switzerland that's believed to harbor the largest concentration of uncontacted peoples on earth. Practically no outsiders are allowed into the preserve. Not scientists. Not anyone. Brazil's National Indian Foundation polices the borders and conducts flyovers to watch for illegal encroachment by loggers.

Protecting the Indians has had the side effect of protecting millions of acres of biodiversity. The Javari Valley remains one of the most species-rich biological hot spots on earth—the rare place where a fish can live in true anonymity. When I realized *that* was the area Heiko had wanted to explore, my head began to spin. I pictured us in a hail of poison-tipped arrows. Or worse, wiping out an entire Indian tribe with a cough. Maybe we wouldn't have gotten far enough upriver to intrude upon the protected reservation. Yet Heiko had told me his goal was to reach the headwaters, where he expected to discover unknown fish species.

For a long time after returning home, I was haunted by the

Jandiatuba, the river that got away. I believed wholeheartedly that the Javari Valley should remain unknown and unknowable. When we touch life, after all, we change it. Yet it can be very hard to keep our hands to ourselves. Every time I caught a glimpse of my *Atlas of the Amazon*, I felt the pang of a missed opportunity—a maddening desire to know what lay hidden in those unexplored waters.

We don't need to wonder any longer. In 2014, Heiko returned yet again to the Amazon and ventured up the Jandiatuba, where he didn't encounter the Arrow People but rather illegal gold prospectors churning up the riverbed. Traveling deep into the river's main tributary, he discovered a wealth of animal life, the likes of which he hadn't seen since he was a child: parrots and herons, frogs and snakes, capybaras and rare giant otters. Of the 181 fish species he recorded, he estimates about 20 to 25 are new to science. The silver arowana was there in small numbers. Colombian poachers had gotten to it first.

HEIKO IS NOT the only one still exploring. "Keep this under your hat, chicken," Tyson Roberts instructed me conspiratorially the last time he called, confessing he planned to go back to Myanmar, "to the land of land mines, spies, and snipers," in search of new species. The 2012 publication of his masterpiece on the world's largest bony fish—*Systematics, Biology, and Distribution of the Species of the Oceanic Oarfish Genus Regalecus*—seemed to give him a new wind. Despite his precarious health, Tyson did eventually return to Myanmar, afterward taking off again across the globe.

Meanwhile, his rival Ralf Britz—an expert on the world's *smallest* known fish, a minnow the size of a Tic Tac—is hot on his heels. Ralf too has traveled back to Myanmar, where he continues to discover new miniature species like the vampire fish, *Danionella dracula*, which has microscopic fangs, and the penis fish, *Danionella priapus*,* the male of which wears its genitals between its pelvic fins. When I eventually met Ralf at the London Natural History Museum, he did not (as Tyson later inquired) have red eyes but rather blue eyes, gray stubble, and an aquiline nose. Like Tyson, he lamented the changing nature of biology, the shift from studying whole animals to analyzing their DNA. "We're definitely losing knowledge," he said. "At a large scale."

In 2014, it was Ralf, posting on a popular aquarium blog, who broke the heartbreaking news that his friend and collaborator of eighteen years—the man without whom I would have gotten nowhere in my quest for the batik arowana—had died at the age of seventy. "Tin Win's deep knowledge of and passion for Myanmar freshwater fishes," Ralf wrote, "made him the single most knowledgeable fish expert in Myanmar." His legacy is kept alive in the spiny eel and golden danio that bear his name, as well as the aquarium-fish exporting company that his son, Hein, and widow, Tin Pyone, continue to run. Even toward the end, when Tin Win knew his heart was failing, he still dreamed of opening a museum to showcase Myanmar's native fishes.

In an age when ichthyology no longer represents the hottest field in science, this zeal for understanding and communing with nature is what I admire most about the aquarium hobby. I am always impressed that serious aquarists know the names of so many species. Linnaeus himself wrote that this was the first step toward wisdom: "For if the name be lost, the knowledge of the thing is also lost." Such dedicated hobbyists are a holdover from the nineteenth century, when amateur naturalists spawned modern biology. Today they pay attention to fish everyone else ignores, notice what's happening to their habitats, and in some

---

* Incidentally, this was the thirtieth fish species to be named after a penis. Only one fish, a goby described in 1801, is named after a vagina—*Trypauchen vagina*.

cases even maintain "arks," sustaining captive populations of species dwindling in nature.

Of course, they exploit fish too, removing them from the wild and plunking them in tanks. Proponents of the hobby argue that the toll is minimal. Globally, perhaps one hundred tons of marine fish are collected as pets each year, whereas 100 million tons are caught as food. Among freshwater fish, habitat destruction is an even greater threat than fishing. Brazil, for example, has approved plans to build more than 150 dams across the Amazon over the next sixteen years to feed the country's growing demand for electricity. Several will be among the largest in the world, including the Belo Monte Dam, being constructed on the Xingu River, where a striking black-and-white catfish called the zebra pleco makes its home. Aquarists eager to breed the species bemoan the irony that it's banned from export while its entire habitat is slated for destruction by 2019. They point out that collection for the aquarium trade has yet to drive a single species extinct. I believe this to be true—at least, I haven't found such a case.

The concept of extinction, however, despite its attractive clarity, may be a misleading measure of what's actually happening to the natural world. The real story of wildlife is told in shades of gray, in the extirpation of populations. In Borneo, I had seen that the Super Red, once a common fish, was now an extreme rarity. In Myanmar, the fisherman in the Tenasserim had reported that the batik arowana was becoming harder and harder to find, requiring travel ever deeper into the jungle. Across the Amazon, I'd heard the same story about the silver arowana and wondered if its depletion was, in turn, affecting the river dolphins and giant otters that eat it.

Increasingly, the silver is now being bred in Asia, which poses its own problems. In Borneo, just outside Sentarum, I visited a village where the locals were keeping the Amazonian species in a pond within reach of the floodwaters. Experts maintain that accidental releases are virtually inevitable in such cases, and the few remaining wild Super Reds may soon be yet another island species forced to compete against a bigger, tougher mainland invasive. Humans have been homogenizing life on land for millennia, spreading the same dominant animals across the globe. Now we are doing so underwater as well, thanks in no small part to the aquarium trade. Many of the seven thousand species the trade moves internationally end up released in local rivers and lakes. The pursuit of novelty has had the paradoxical consequence of blending everything together, which inevitably results in a stultifying sameness.

Despite our endless desire for the next new thing, Kenny the Fish appears to have been right: the dragon fish has—at least for now—proven its staying power as a pet. From New York to Shanghai, arofanatics continue to lust after the prehistoric creature, their passion driving them to unlikely extremes. In 2015, the man who is probably the world expert on the Asian species—Alex Chang, Kenny the Fish's head research scientist—was arrested in Australia for trying to smuggle twenty-some Asian arowana (among other fish) into the country in his luggage.

Reading about Alex's arrest in the news, I thought back to my first day reporting in Singapore when the cheerful scientist was the first to show me an arowana in person, gushing about how the fish looked like a solid gold bar. Since then, I'd always turned to him with any questions about the biology of the species. Alex had spent six years studying its reproduction and genetics to earn his doctorate. Now he was facing up to a decade in an Australian prison.* I couldn't wrap my head around it.

Yet even I had trouble disentangling myself from the grip of the arowana, which had come to dominate my life. Despite

---

* In November 2015, after spending several months behind bars and the better part of a year under house arrest, Alex received a sentence of a year and nine months in prison, suspended based on time served.

finding the object of my quest, I could not, in fact, stop questing. When Heiko made plans to return to the Amazon and travel up the Jandiatuba, he invited me to join him on that expedition. Despite myself, I gave the idea serious consideration, until Jeff finally put his foot down, insisting my obsession had gone too far. So I stopped.

Now we are expecting a baby—a Pisces. Someday my child may want a pet fish. For my part, I think Stephen will remain my last. The memory of his gasping SpaghettiO of a mouth still makes me cringe. I did, however, purchase a Wardian case, the predecessor to the aquarium, which the London doctor Nathaniel Bagshaw Ward invented in 1829. About a foot tall, it looks like a little glass greenhouse and is perfectly sealed so moisture can't escape, creating a self-sustaining ecosystem. Inside I planted a fern, which is thriving despite my not having watered it in a year. Every so often, I notice this dewy cube of greenery in my small city apartment and am transported back to a wild place.

Acknowledgments

This book came to be because Lieutenant John Fitzpatrick was a natural-born storyteller. I called him out of the blue one summer afternoon and found I could scarcely take notes fast enough. I didn't know anything about wildlife trafficking back then. I didn't care one iota about fish. Tragically, in May 2014, John died of heart failure at forty-six, leaving behind his beloved wife and three-month-old daughter. I feel fortunate to have captured one morning in the life he lived so passionately. I will always be grateful for the grand adventure he set me on.

I could not have undertaken such an ambitious project without the generous support of a Pulitzer Traveling Fellowship. I thank the Columbia University Graduate School of Journalism, especially Marguerite Holloway, Arlene Morgan, and John Bennet, whose voice I still hear in my head when revising. I am indebted to Jonathan Weiner for inspiring me to pursue science writing in the first place.

My phenomenal agent, Abigail Koons, has nurtured this project since it was the seed of an idea, offering continuous encouragement and sage feedback through the years, cheering me on to the finish line. At Scribner, I thank Colin Harrison for his enduring support, Katrina Diaz for her inspired editing, Sarah Goldberg for taking on the project with gusto, Nan Graham for her initial encouragement, and Paul Whitlatch for his great enthusiasm and editorial insights. I am grateful to the sharp-eyed Steve Boldt for his meticulous and thoughtful copyediting. Kyle Kabel designed the elegant interior and Na Kim the beautiful cover. My thanks to Jessica Yu, Angela Baggetta, and Kathleen Zrelak for their dedicated publicity efforts, and Kara Watson and Ashley Gilliam for their work on marketing.

Beth Rashbaum provided truly invaluable editorial development, especially in streamlining a complex narrative structure. Jessica Seigel improved the book tremendously with her brilliant critical eye, acting as a pillar of strength through the life of the project and especially in the final stretch. Writing guru Hillary Rettig was a great support in the early stages of my work and continues to be a good friend. I also thank Alexandra Shelley and Nancy Rawlinson for their valuable insights. Jad Abumrad of *Radiolab* initially encouraged me to pursue a story on the exotic pet trade—though I never did find him that wildebeest in Queens.

I am deeply grateful to Kelly Caldwell, Liesl Schwabe, and Stephanie Paterik, who provided not only camaraderie and laughter but tireless insight on countless drafts. Laura Castellano Richards, who has been reading about Kenny the Fish since the day I first clicked on his belly button, greatly improved the book with her feedback and kept me afloat with her friendship. Brooke Borel offered keen perspectives on early chapters. Neuwrite 2 served as a helpful forum in which to vet much of the science.

When it came time for fact-checking, Karen Fragala-Smith tackled the sizeable task with an impressive can-do spirit that made the process less daunting. Brad Scriber also contributed to the review, taking on sections particularly heavy on science and history. Mark Szorc translated excerpts of Amanda Bleher's diaries from the original German and provided historic expertise on crucial matters such as toilet paper from the post–World War II era. Kristen Brown carefully transcribed many hours of interviews.

For inspiring me to write this book and tolerating my presence through many weeks of reporting, I am grateful to the extraordinary Heiko Bleher, as well as his wife, Natasha Khardina, for her support. At the start of this project, I did not know I would cross paths with Tyson Roberts and Ralf Britz, and I could not have been more fortunate on both accounts. I am indebted to each of them for taking the time to share with me the deep passion that drives great scientists. For ushering me into the world of the arowana, I thank the inimitable Kenny the Fish and the dragon whisperer Alex Chang.

Numerous other people generously gave of their time and knowledge throughout my reporting, including: Jake Adams, Syed Abas Alattas, Glen Axelrod, Eddie Badaruddin ("Mr. Bad"), Sahibol Anwar Bin Arba'e, Ahmad Bin Don, A. Cuneyt Birol, Binti Brindamour, John Carberry, Chan Weng Kei, Frank Chang, Simon Chaw, Li-Wei Chih ("Neoprodigy"), Vincent Chong, Keith Davenport, John Dawes, Bernd Degen, Scott Dowd, Albert Ee, Hans-Georg Evers, Takehito Fuikui, Mark Gardner, Richard Garriott, Steve Gibbons, Dave Goh, Richard Goh, Eben Haezer, Hariffin, Hermanto, Hermanus Haryanto, Hlaing Win, Hla Win, Esteban Meléndez Holanda, Hsu Chiao Chen, En Ibrahim, Daniel Indarta, Pa Itam, Jap Khiat Bun, Lorie Karnath, Cedric Koh, Pa Juniardi, Eugene Jussek, Tsuyoshi Kawakami, Wilford Landong, Larry Law, Patrick Law, Terry Law, A.J. Lee, William Lee, Hendri Leong, Lin Jieming, Ling Kai Huat, Steven Lundblad, Kate McGill, Andy Moo, Sandy Moore, Ester Mous, Myint Too, Nahary Latifah, Michael Neo, Ng Huan Tong, Max Ng, John Niemans, Manabu Ogata, Raka Dwi Permana, Russell Peters, Phelisonia, Eko B. Priyongggo, Michael Salter, Jim Sambi, Doyle Schafer, Willie Si ("Dr. Arowana"), Andrew Soh, Soe Nyunt Tun, Julian Sprung, Su Wen

Hung, Stan Sung, Stephen Suryaatmadja, Suwandi, Yasumasa Takahashi, Admond Tan, Tan Cheng Kiat, Kevin Tan, Jimmy Tan, Tan Lai Soon, Tan Nam Wah, Tony Tan, Tris Tanoto, Tin Tin Aye, Michael Toh, Kamphol Udomritthiruj, Henny van Groesen, Don Walsh, Masahiro Yamamoto, Satoru Yamamoto, Desmond Yeoh, Yamazaki Daiya, Yamazaki Yoshiyuki, Alvin Yap, Andy Yap, Yap Kim Choon, Amos Yu, Humberto Zea, and Pedro Zea. I am particularly indebted to Martin Toh for ushering me into the world of arofanatics; Hery Cheng and Willy Sutopo for hosting me in Borneo; U Tin Win, Daw Tin Pyone, and Ko Ye Hein Htet for their kind help in Myanmar; Khor Harn Sheng for driving me through Malaysia and acting as an impromptu interpreter; Andrew Lim for helping facilitate my reporting; Raymond Cheah for welcoming me into his home; Tor Kreutzman for his good humor and perspectives; and Stephanie Lee and Osmond Chao for their company in Taiwan.

Throughout this project, I was fortunate to have the ear of Svein Fosså, who provided keen perspectives not only on the trade but also the larger philosophical issues surrounding the hobby. At Ornamental Fish International, Gerald Bassleer and Alex Ploeg both offered critical insights. In July 2014, I learned with great sorrow that Ploeg, along with his wife, son, and friend, died aboard Malaysia Airlines Flight 17, shot down over Ukraine. His kind spirit and contagious laugh are deeply missed.

A number of scientists were especially helpful to me, including: Carol Colfer of the Center for International Forestry Research, Oliver Coomes of McGill University; Frank Courchamp of the University of Paris South; Walter Courtenay of Florida Atlantic University; Lisa Curran of Stanford University; Richard Dudley of Cornell University; Fabrice Duponchelle of the Institute of Research for Development, Marseilles; Mark Erdmann of Conservation International; Carl Ferraris of the California Academy of Sciences; Maurice Kottelat, president of the European Ichthyological Society; Sven O. Kullander of the

Swedish Museum of Natural History; John Long of Flinders University; Camilo Mora of the University of Hawaii; Ng Heok Hee, Peter Ng, and Tan Heok Hui of the National University of Singapore; László Orbán of Temasek Life Sciences Laboratory; Theodore Pietsch of the University of Washington; Fabrice Teletchea of University of Lorraine; and Mohd Zakaria-Ismail of the University of Malaya. In the field of conservation, I am particularly grateful to Chris Shepherd and Sabri Zain of TRAFFIC International; Heri Valentinus of Riak Bumi; Tom de Meulenaer, Vincent Fleming, and David Morgan of CITES; Steve Edwards, formerly of the IUCN; Grahame Webb of Wildlife Management International; Johanna Fischer of the Food and Agriculture Organization of the United Nations; Teresa Telecky of Humane Society International; Sandre Altherr of Pro Wildlife; D. J. Schubert of the Animal Welfare Institute; and Colman O'Criodain, Yuyun Kurniawan, Albertus Tjiu, Adam Tomasek, U Tin Than, and Elizabeth Wetik of the World Wildlife Fund. Marshall Meyers of the Pet Industry Joint Advisory Council provided a wealth of knowledge on the legal landscape of the wildlife trade, as did Paul Chapelle of the US Fish and Wildlife Service, Craig Hoover and Rick Parsons, formerly of USFWS, and Major Scott Florence of the New York Department of Environmental Conservation.

Above all, I am grateful to my family. As Hein once told me of Tin Win, "Without his sweet love, I would not be so success." I feel exactly the same about my mother and father, Julie and Henry Voigt, who not only provided important feedback on the manuscript and practical advice during my reporting, but have buoyed me with great love throughout my life. I am fortunate to have gained a second pair of equally loving parents through marriage, Susan and David Korn, who encouraged me in this project every step of the way. My husband, Jeff Korn, has been my single greatest, unwavering support since the night I first told him, "So there's this fish . . ."

Notes on Sources

Many, many people—far more than appear in this book—helped me understand the story of the arowana and the cultures surrounding it. Others gave me a crash course in ichthyology and the history of natural science. A single shelf of fish books grew into an entire bookcase, which is now prone to avalanches when bumped. Though I regret not being able to include all the people, books, and studies that have contributed to my research, I list my major sources below.

Prologue

I am grateful to the Chan family for sharing the story of their son Chan Kok Kuan. I could not have found them without the dogged footwork of Danny Lim and Tong Yee Siong in Malaysia—the only instance in this book where distance and language barriers necessitated my reliance on the work of local journalists. Tong's translated interviews and Lim's documentation, recordings, and photographs made it possible for me to re-create the events of May 11, 2004.

PART I
"DRAGONS IN THEIR PLEASANT PALACES"

Chapter One: The Pet Detective

For the story of Richard Ogust and his twelve hundred turtles, see the 2006 PBS documentary *The Chances of the World Changing*. In tracing the human obsession with exotic animals through history, I have relied on numerous sources, including *Looking at Animals in Human History* by Linda Kalof (London: Reaktion Books, 2007); *In the Company of Animals: A Study of Human-Animal Relationships* by James Serpell (Oxford: B. Blackwell, 1986); *Dominance & Affection: The Making of Pets* by Yi-Fu Tuan (New Haven: Yale University Press, 1984); *Pets in America: A History* by Katherine C. Grier (Chapel Hill: University of North Carolina Press, 2006); and *Companion Animals in Society* by Stephen Zawistowski (Clifton Park, NY: Thomson/Delmar Learning, 2008). James Serpell helped make sense of the data available on pet ownership in America. The statistics I cite come from the 2015–16 National Pet Owners Survey conducted by the American Pet Products Manufacturers Association. I first read that more exotic pets are believed to live in American homes than zoos in Lauren Slater's "Wild Obsession: The perilous attraction of owning exotic pets," *National Geographic* (April 2014). The journalist Bryan Christy has done extensive reporting on the illegal wildlife trade, concluding that it may well be the most profitable form of illegal trade, bar none.

Arowana taxonomy is no simple matter, and I thank Carl Ferraris for talking me through its complexities. It has long been acknowledged that the South American arapaima and the African arowana are more closely related to each other than to the rest of the arowanas. In *Check List of the Freshwater Fishes of South and Central America* (Porto Alegre, Brazil: EDIPUCRS, 2003), Ferraris, Roberto E. Reis, and Sven O. Kullander place these two species in their own separate family of Arapaimatidae—a position now adopted by the *Catalog of Fishes* edited by William N. Eschmeyer and Ronald Fricke (San Francisco: California Academy of Sciences) and available online at http://researcharchive .calacademy.org/research/ichthyology/catalog/fishcatmain.asp. Other sources, including the fourth edition of *Fishes of the World* by Joseph S. Nelson (Hoboken, NJ: John Wiley & Sons, 2006), continue to classify all the arowanas and the arapaima together in the family Osteoglossidae. (Complicating matters, Nelson also includes the African butterflyfish in the family.) Here and in later chapters, my go-to source on fish biology has been the second edition of *The Diversity of Fishes: Biology, Evolution, and Ecology* by Gene S. Helfman, Bruce B. Collette, Douglas E. Facey, and Brian W. Bowen (Hoboken, NJ: Wiley-Blackwell, 2009).

For the articles about the arowana I refer to, see in the *Straits Times* of Singapore "High-security fish farm is big business" (November 20, 2004), "Yet another fish theft—seven arowana lost" (March 24, 2002), "Jail, cane for bid to steal fish" (April 15, 2005), "Stolen catch" (July 21, 2002), "Learn from the fishes" (March 22, 2002), and "Who can be Singapore's Donald Trump?" (March 16, 2008); "Therapeutic hobby of keeping fish," the *New Straits Times* of Malaysia (February 6, 1995); "Arowana become popular pet of super rich," *Practical Fishkeeping* (March 25, 2008); "Importer kidnapped for 11 million yen ransom in Jakarta," the *Daily Yomiuri* of Japan (January 13, 1999); "Fishy noise hides undercover job," the *Herald Sun* of Australia (June 7, 2005); and "Qian Hu: A bigger fish than meets the eye?," the *Business Times* of Singapore (April 4, 2002).

Chapter Two: The Fish

Edwin Thumboo's poem "Ulysses by the Merlion" is published in a collection by the same name (Singapore: Heinemann Educational Books, 1979). For the range of the Asian arowana, I have relied on data from Ng Heok Hee of the University of Singapore and Maurice Kottelat of the European Ichthyological Society, as well as the IUCN Red List of Threatened Species. A study of the impact of trade on the wild fish in Cambodia is "Harvest, trade and conservation of the Asian arowana *Scleropages formosus* in Cambodia" by Jodi J. L. Rowley, David A. Emmett, and Seila Voen, *Aquatic Conservation: Marine and Freshwater Ecosystems* 18 (November/December 2008). My information on the global distribution of arowanas comes from Tim M. Berra, *Freshwater Fish Distribution* (Chicago: University of Chicago Press, 2007).

Tan Heok Hui of the Raffles Museum of Biodiversity Research at the National University of Singapore provided much insight into the world of ichthyology as well as access to historical materials. I'm grateful to collections manager Kelvin Lim for digging up the 1969 paper "Conserving Malayan fresh-water fishes" by Eric R. Alfred, *Malayan Nature Journal* 22 (1969). My account of the introduction of the fish to the aquarium trade is based on an in-person interview with Chew Thean Yeang at his fish shop in Penang, Malaysia. For a detailed history of Singapore's early breeding of the Asian arowana, see *The Dragon Fish* by John Dawes, Lim Lian Chuan, and Leslie Cheong (Waterlooville: Kingdom Books, 1999). Statistics on Singapore's aquarium exports are courtesy of the Singapore Agri-Food & Veterinary Authority.

Chapter Three: The Arowana Cartel

All statistics relating to CITES are available through the convention's website, cites.org/eng, which includes a trade database and registry of captive-breeding facilities. For information on palm oil, see the website of the World Wildlife Fund, particularly worldwildlife.org/pages/which-everyday-products-contain-palm-oil.

The estimate that the aquarium trade moves 7,000 species comes from adding together the 5,325 freshwater species listed in *Standard Names for Freshwater Fishes in the Ornamental Aquatic Industry* by Roberto R. Hensen, Alex Ploeg, and Svein A. Fosså (Maarssen, the Netherlands: Ornamental Fish International, 2010) and the 1,802 marine species cited in "Revealing the appetite of the marine aquarium fish trade; the volume and biodiversity of fish imported into the United States" by Andrew L. Rhyne, Michael F. Tlusty, Pamela J. Schofield, Les Kaufman, James A. Morris Jr., and Andrew W. Bruckner, *PLOS ONE* 7 (May 21, 2012). See also: "The Volume of the Ornamental Fish Trade" by Alex Ploeg in *International Transport of Live Fish in the Ornamental Aquatic Industry* edited by Svein Fosså, Gerald Bassleer, and Alex Ploeg (Maarssen, the Netherlands: Ornamental Fish International, 2007). The estimate that 90 percent of freshwater ornamental fish are now farmed comes from the Food and Agriculture Organization (FAO) of the United Nations (fao.org/fishery/topic/13611/en). Further sources on the aquarium trade include "Uncovering an obscure trade: Threatened freshwater fishes and the aquarium pet markets" by Rajeev Raghavan,

Neelesh Dahanukar, Michael F. Tlusty, Andrew L. Rhyne, K. Krishna Kumar, Sanjar Molur, and Alison M. Rosser, *Biological Conservation* 164 (2013), and "The benefits and risks of aquacultural production for the aquarium trade" by Michael Tlusty, *Aquaculture* 205 (2002).

The death threat against Malaysian politician Wee Choo Keong was reported in "Wee lodges report on 'death threat,' " the *Star Online* (April 7, 2011). For a thorough history of the Chinese settlement of West Borneo and the role of *kongsi* in the region, see Mary Somers Heidhues's *Golddiggers, Farmers, and Traders in the "Chinese Districts" of West Kalimantan, Indonesia* (Ithaca, NY: Southeast Asia Program Publications, Cornell University, 2003).

Chapter Four: Aquarama

Julio Cortázar's "Axolotl" appears in his short-story collection *Final del Juego* (Buenos Aires: Editorial Sudamericana, 1964). The value of global exports of ornamental fish from 1976 to the present comes from the FAO. The valuation of the industry varies from $15 billion to $30 billion. I cite the more conservative figure based on the wisdom of Keith Davenport of the Ornamental Aquatic Trade Association.

For an authoritative study of fish-keeping in ancient Rome, see James Higginbotham's *Piscinae: Artificial Fishponds in Roman Italy* (Chapel Hill, NC: University of North Carolina Press, 1997). The invention of the aquarium is recounted in Lynn Barber's illuminating book *The Heyday of Natural History, 1820–1870* (Garden City, NY: Doubleday, 1980), which pointed me to the works of Philip Henry Gosse, who popularized the hobby in the 1850s. Also helpful was Stephen Jay Gould, "Seeing Eye to Eye, Through a Glass Clearly," *Natural History* 106 (July/August 1997), Bernd Brunner's *The Ocean at Home: An Illustrated History of the Aquarium* (New York: Princeton Architectural Press, 2005), and Albert Klee's *The Toy Fish: A History of the Aquarium Hobby in America—the First One Hundred Years* (Pascoag, RI: Finley Aquatic Books, 2003). The latter pointed me to "The Myth of the Balanced Aquarium," by James W. Atz, *Natural History* 58 (1949). For an enlightening account of the history of ecology, and the pervasive myth of a steady state in nature, see Donald Worster's *Nature's Economy: A History of Ecological Ideas*, 2nd ed. (New York: Cambridge University Press, 1994).

My primary source on Heiko Bleher's childhood and the life of his mother, Amanda Bleher, is her memoir, *Iténez: River of Hope* (Miradolo Terme: Aquapress, 2009), published posthumously more than thirty years after she wrote it. To confirm what I could of the early history of the Bleher family, I spent much time poring through Amanda's papers and diaries, as well as newspaper clippings about her fish-and-aquatic-plant exporting business and that of her father, Adolf Kiel. I spoke with Heiko's elder sister Irene Bleher, and in 2012, I visited Amanda's close friend Eugene Jussek, whom she first met in Frankfurt during World War II and with whom she maintained a lifelong correspondence. A psychiatrist in Los Angeles specializing in hypnotherapy and past lives, Jussek recalled dancing with Amanda to the smoky voice of Zarah Leander and was able to provide an invaluable perspective on her character and life.

Chapter Five: The Dragon's Den

My information on Pekingese court dogs comes from James Serpell's *In the Company of Animals: A Study of Human-Animal Relationships* (Oxford: B. Blackwell, 1986). The quotation from the priest in Borneo appears in Richard Lloyd Parry's account of the ethnic violence between the Dayak and the Madurese, *In the Time of Madness: Indonesia on the Edge of Chaos* (London: Jonathan Cape, 2005). The statistics on the logging of Borneo are courtesy of Lisa Curran of Stanford University who has spent nearly forty years studying the island. The growth of the oil palm industry is quantified in "How will oil palm expansion affect biodiversity?" by Emily B. Fitzherbert, Matthew J. Struebig, Alexandra Morel, Finn Danielsen, Carsten A. Brül, Paul F. Donald, and Ben Phalan, *Trends in Ecology and Evolution* 23 (October 2008).

PART II

THE EDGE OF THE KNOWN WORLD

Chapter Six: The Living Fish

Heiko Bleher has written a series of informative articles on the history of fishkeeping in *Nutrafin Aquatic News*, issues 1–4. The story of the

Spanish farmer who discovered the "Great Fish" is recounted in José Bullón Giménez's *The Pileta Cave: National Monument since 1924* (Alcalá del Valle, Spain: La Serranía, 2006). John Milton's description of the Assyrian fish god Dagon appears in lines 462–63 of *Paradise Lost* (1667). The information on fish mummies in ancient Egypt comes from a conversation with Salima Ikram of the American University of Cairo and her book *Divine Creatures: Animal Mummies in Ancient Egypt* (New York: American University in Cairo Press, 2005).

In writing about Aristotle's biology, I have relied not only on his own work *History of Animals*, but also the 2011 BBC documentary *Aristotle's Lagoon* hosted by Armand Marie Leroi, who has since written a book on the subject: *The Lagoon: How Aristotle Invented Science* (New York: Bloomsbury Circus, 2014). Another valuable resource was James Lennox, "Aristotle's Biology," *The Stanford Encyclopedia of Philosophy* (Spring 2014 edition), edited by Edward N. Zalta (http://plato.stanford.edu/archives/spr2014/entries/aristotle-biology/). Lennox's translation of the "Invitation to Biology" appears in *Aristotle: On the Parts of Animals* (Oxford: Oxford University Press, 2001).

My characterization of Peter Artedi owes much to Theodore W. Pietsch's historical novel *The Curious Death of Peter Artedi: A Mystery in the History of Science* (New York: Scott & Nix, 2010), which squeezes every drop out of the sparse primary source materials available on Artedi. I am grateful to Pietsch for speaking with me about his extensive research into Artedi's influence on Linnaeus. Two papers were also of great help to me: "Peter Artedi, founder of modern ichthyology" by Alwyne Wheeler, who also suggests that Artedi's full contribution to biology has been obscured, and Gunnar Broberg's "Petrus Artedi in his Swedish context," both published in the *Proceedings of the Fifth Congress of European Ichthyologists, Department of Vertebrate Zoology*, edited by Sven Kullander and Bo Fernholm (Stockholm: Swedish Museum of Natural History, 1987).

My understanding of the great age of natural exploration is rooted in many sources, but I found two particularly enlightening: Paul Lawrence Farber's *Finding Order in Nature: The Naturalist Tradition from Linnaeus to E. O. Wilson* (Baltimore: Johns Hopkins University Press, 2000) and Richard Conniff's *The Species Seekers:*

Heroes, Fools, and the Mad Pursuit of Life on Earth (New York, W. W. Norton, 2011).

Chapter Seven: The Explorers

As mentioned in the notes for chapter 4, information about Amanda Bleher comes from her own writings and papers as well as from the recollections of those who knew her. Tony Juniper, author of *Spix's Macaw: The Race to Save the World's Rarest Bird* (New York, Washington Square Press, 2002), provided an expert perspective on the history of the species and a counterpoint to the idea that the CITES listing contributed to its demise. Wade Davis's study of Haitian "zombies" is detailed in *The Serpent and the Rainbow: A Harvard Scientist's Astonishing Journey into the Secret Society of Haitian Voodoo, Zombis, and Magic* (New York: Simon & Schuster, 1985).

In writing about Edward O. Wilson, I have relied on his autobiography, *Naturalist* (Washington, DC: Island Press, 1994), and *The Future of Life* (New York: Alfred A. Knopf, 2002). Global extinction rates are based on mathematical modeling and much debated. Those I cite are the latest espoused by Wilson. For a full account of Wilson's biophilia hypothesis, see his book *Biophilia* (Cambridge, MA: Harvard University Press, 1984) and the anthology *The Biophilia Hypothesis* (Washington, DC: Island Press, 1993) edited by Stephen R. Kellert and Wilson. The latter details two studies regarding the physiological effects of watching fish in aquariums: "Looking, talking and blood pressure: The physiological consequences of interaction with the living environment" by Aaron Katcher, E. Friedmann, A. Beck, and J. Lynch, *New Perspectives on Our Lives with Companion Animals*, edited by Katcher and Beck (Philadelphia: University of Pennsylvania Press, 1983); and "Comparison of contemplation and hypnosis for the reduction of anxiety and discomfort during dental surgery" by Katcher, H. Segal, and Beck, *American Journal of Clinical Hypnosis* 1 (1984). More recently, similar results have been reported in "Marine biota and psychological well-being: A preliminary examination of dose-response effects in an aquarium setting" by Deborah Cracknell, Mathew P. White, Sabine Pahl, Wallace J. Nichols, and Michael H. Depledge, *Environment and Behavior* (July 28, 2015).

For more on Eugenie Clark, see her autobiography, *The Lady and the Sharks* (New York: Harper & Row, 1969). The estimated annual slaughter of sharks comes from "Global estimates of shark catches using trade records from commercial markets" by Shelley C. Clarke, Murdoch K. McAllister, E. J. Milner-Gulland, G. P. Kirkwood, Catherine G. J. Michielsens, David J. Agnew, Ellen K. Pikitch, Hideki Nakano, and Mahmood S. Shivji, *Ecology Letters* 9 (October 2006).

Chapter Eight: Naming Rights

Much debate exists not only about the total number of species on earth but even about how many species have already been described, depending on the sources counted and methods used to avoid redundancies. I have relied on the second edition of Arthur Chapman's "Numbers of Living Species in Australia and the World" for the Australian Biological Resources Study (September 2009). See also: "Global species richness estimates have not converged" by M. Julian Caley, Rebecca Fisher, and Kerrie Mengersen, *Trends in Ecology & Evolution* 29 (April 2014); and "How many species are there on earth and in the ocean?" by Camilo Mora, Derek P. Tittensor, Sina Adl, Alastair G. B. Simpson, and Boris Worm, *PLOS Biology* 9 (April 2011). Mora and Caley both helped me make sense of the data.

Melanie Stiassny's quotation comes from her 1997 presentation with Ian J. Harrison, "Vanishing from freshwater: Species decline and the machinery of extinction," at the American Museum of Natural History (amnh.org/science/biodiversity/extinction/Day1/bytes/Stiassny Pres.html). The average number of fish species discovered each year is based on data from the *Catalog of Fishes* (see chapter 1).

The paper dividing the Asian arowana into multiple species is "The different colour varieties of the Asian arowana *Scleropages formosus* (Osteoglossidae) are distinct species: Morphologic and genetic evidences" by Laurent Pouyaud, Tomy Sudarto, and Guy G. Teugels, *Cybium* 27 (2003). Although Pouyaud did not respond to requests for an interview, I was able to meet with his coauthor Sudarto in Jakarta and Mark Erdmann, discoverer of the Indonesian coelacanth, in Bali. For more on the controversy over the Indonesian coelacanth, see "Tangled tale of a lost, stolen and disputed coelacanth," *Nature* (July 13, 2000) by Heather McCabe and Janet Wright. Samantha Weinberg's *A Fish Caught in Time: The Search for the Coelacanth* (New York: Harper-Collins, 2000) provides a fascinating account of the full story of the coelacanth. For the study sequencing the entire coelacanth genome, see "The African coelacanth genome provides insights into tetrapod evolution" by a consortium of international experts, *Nature* 496 (April 18, 2013).

The growing number of species in the aquarium trade comes from Svein A. Fosså's illuminating article "Man-made fish: Domesticated fishes and their place in the aquatic trade and hobby," *OFI Journal* 44 (February 2004). The information on Herbert Axelrod appears in "The strange fish and stranger times of Dr. Herbert R. Axelrod," *Sports Illustrated* (May 3, 1965); "A life of money and myths," *Star-Ledger* (April 25, 2004); and *Discus* by Bernd Degen and Herbert Axelrod (Ruhmannsfelden, Germany: Degen Mediahouse, 2011). I am also grateful to Herbert Axelrod's nephew Glen Axelrod and Gary Hirsch for their perspectives.

PART III

THE SUPER RED

Chapter Nine: In the Age of Aquariums

I first came across Ida Pfeiffer's name in "The fishes of Danau Sentarum National Park and the Kapuas Lakes Area, Kalimantan Barat, Indonesia" by Maurice Kottelat and Enis Widjanarti, *Raffles Bulletin of Zoology* 13 (2005). The best source on Ida Pfeiffer is Ida herself, particularly her books *A Woman's Journey Round the World* (London: Peter Duff & Co, 1852) (for the sake of consistency with the sequel, I've translated the title as *A Lady's Journey Round the World*); *A Lady's Second Journey Round the World* (New York: Harper & Brothers, 1856); and *The Last Travels of Ida Pfeiffer: Inclusive of a visit to Madagascar, with a biographical memoir of the author* (New York: Harper & Brothers,1861). For a modern perspective on Ida's travels, see "Woman on the road: Ida Pfeiffer in the Indies" by Mary Somers Heidhues, *Archipel* 68 (2004).

To tell the story of Alfred Russel Wallace, I've relied on his own works *The Malay Archipelago* (New York: Harper & Brothers, 1869) and *My Life: A Record of Events and Opinions* (New York: Dodd, Mead & Company, 1905), as well as numerous secondary sources, including Peter Raby's *Alfred Russel Wallace: A Life* (Princeton, NJ: Princeton University Press, 2001); Loren Eiseley's *Darwin and the Mysterious Mr. X: New Light on the Evolutionists* (New York: Dutton, 1979); David Quammen's *Song of the Dodo: Island Biogeography in an Age of Extinction* (New York: Scribner, 1996); Jonathan Rosen's "Missing link: Alfred Russel Wallace, Charles Darwin's neglected double," the *New Yorker* (February 12, 2007), and Tony Whitten's introduction to *The Malay Archipelago* (Singapore: Periplus Editions, 2008).

The study solving the century-old mystery of how the arowana ended up in Southeast Asia is Yoshinori Kumazawa and Musumi Nishida's "Molecular phylogeny of Osteoglossoids: A new model for Gondwanian origin and plate tectonic transportation of the Asian arowana," *Molecular Biology and Evolution* 17 (2000). A fascinating account of fish evolution, detailing how the bonytongues came to dominate the seas, is John A. Long's *The Rise of Fishes: 500 Million Years of Evolution* (Baltimore: Johns Hopkins University Press, 2011).

Statistics on the species richness of tropical rain forests come from the Edward O. Wilson's *The Future of Life* (New York: Alfred A. Knopf, 2002). For rates of deforestation in Borneo, see *Borneo: Treasure Island at Risk* (Frankfurt am Main: WWF Germany, 2005). The statistics on freshwater fish diversity are recounted in Stiassny and Harrison (see chapter 8). The data on species extinctions in Singapore appear in Barry W. Brook, Navjot S. Sodhi, and Peter K. L. Ng's "Catastrophic extinctions follow deforestation in Singapore," *Nature* 424 (July 24, 2003). Also helpful to me was the fifth edition of Richard B. Primack's *Essentials of Conservation Biology* (Sunderland, MA: Sinauer Associates, 2010).

Chapter Ten: Ghost Fish

Wim Giesen's in-depth reports on the lake region: *Danau Sentarum Wildlife Reserve: Inventory, Ecology and Management Guidelines* for the World Wildlife Fund (1987) and *Habitat Types and Their Management: Danau Sentarum* for Wetlands International (1996) were invaluable in characterizing the ecology of Sentarum. A number of papers published in the *Borneo Research Bulletin* 41 (2010) also provided critical background, including "Danau Sentarum National Park, Indonesia: A historical overview" by Julia Aglionby, "Fluid landscapes and contested boundaries in Danau Sentarum" by Emily E. Harwell; and "Interacting threats and challenges in protecting Danau Sentarum" by Valentinus Heri, Elizabeth Linda Yuliani, and Yayan Indriatmoko. Conversations with anthropologist Carol Colfer and fisheries biologist Richard Dudley were a great help. Valentinus Heri, a founder of Riak Bumi, provided invaluable perspectives on life in Sentarum and kindly helped arrange my trip.

As above, the material on Ida Pfeiffer is primarily from her own writings. For the story behind the original type specimens of the Asian arowana, see "The types of *Osteoglossum formosum* Müller & Schlegel, 1840 (Teleostei, Osteoglossidae)" by Martien J. P. Van Oijen and Sancia E. T. Van Der Meij, *Zootaxa* 3722 (2013). I discovered the shorthead hairfin anchovy in Pieter Bleeker's papers at the American Philosophical Society in Philadelphia. The passage from Alfred Russel Wallace appears in his classic travelogue *The Malay Archipelago* (New York: Harper & Bros, 1869).

PART IV

FISH #32,107

Chapter Eleven: Monsters of Our Making

"The dangerously venomous snakes of Myanmar" by Alan E. Leviton, Guinevere O. U. Wogan, Michelle S. Koo, George R. Zug, Rhonda S. Lucas, and Jens V. Vindum appeared in the *Proceedings of the California Academy of Sciences* 54 (November 14, 2003). For a full account of the tragic death of Joe Slowinski, see "Bit" by Mark W. Moffett, *Outside Magazine* (January 4, 2002).

The challenge of defining a fish is nicely characterized on page 3 of Helfman et al. (see chapter 1). Charles Darwin's writings on goldfish appear in his book *The Variation of Animals and Plants Under Domestication: Volume I* (London: John Murray, 1868). For a collection of everything Darwin ever wrote about fish, see *Darwin's Fishes: An*

Encyclopedia of Ichthyology, Ecology, and Evolution by Daniel Pauly (New York: Cambridge University Press, 2004).

On the subject of domestication, I have relied on numerous sources, including Juliet Clutton-Brock's *Animals as Domesticates: A World View through History* (East Lansing: Michigan State University Press, 2012), particularly the foreword by James A. Serpell, which pointed me to the estimated biomass of domesticated vertebrates in Vaclav Smil's *General Energetics: Energy in the Biosphere and Civilization* (New York: Wiley, 1993); the second edition of *Genetics and the Behavior of Domestic Animals* edited by Temple Grandin and Mark J. Deesing (London: Academic Press, 2014); and *Domestication* by Clive Roots (Wesport, CT: Greenwood Press, 2007).

Fabrice Teletchea of the University of Lorraine helped me find my bearings in the poorly studied field of fish domestication. Teletchea and Pascal Fontaine argue that the 250 species reared in ponds now lie on the domestication spectrum in "Levels of domestication in fish: Implications for the sustainable future of aquaculture," *Fish and Fisheries* 15 (2014). For the story of the common carp, I have also relied on E. K. Balon's "About the oldest domesticates among fishes," *Journal of Fish Biology* 65 (2004); and Svein A. Fosså's "Man-made fish: Domesticated fishes and their place in the aquatic trade and hobby," *OFI Journal* (February 2004), which provides the historical count of domesticated fishes in the hobby. For a detailed history of koi, see Michugo Tamadachi's *The Cult of the Koi* (Neptune City, NJ: TFH Publications, 1990).

The data on the growth of aquaculture comes from the FAO. The behavior of domesticated fish and their interactions with wild fish are detailed in the second edition of Rex A. Dunham's *Aquaculture and Fisheries Biotechnology: Genetic Approaches* (Wallingford: CABI, 2011). For the effects of interbreeding between wild and domesticated salmon, see, for example, "Fitness reduction and potential extinction of wild populations of Atlantic salmon, *Salmo salar*, as a result of interactions with escaped farm salmon" by Philip McGinnity, Paulo Prodöhl, Andy Ferguson, Rosaleen Hynes, Niall ó Maoiléidigh, Natalie Baker, Deirdre Cotter, Brendan O'Hea, Declan Cooke, Ger Rogan, John Taggart, and Tom Cross, *Proceedings of the Royal Society B* 270 (December 7, 2003).

Chapter Twelve: "The Authorities Will Be Watching You"

The *Catalog of Fishes* (see chapter 1) keeps a running tally of known fish species. Numerous sources provided an essential primer in Burmese history, including Emma Larkin's *Finding George Orwell in Burma* (New York: Penguin Press, 2004) and *Everything Is Broken: A Tale of Catastrophe in Burma* (New York: Penguin Press, 2010); and Thant Myint-U's *River of Lost Footsteps: A Personal History of Burma* (New York: Farrar, Straus and Giroux, 2006) and *Where China Meets India: Burma and the New Crossroads of Asia* (New York: Farrar, Straus and Giroux, 2011).

The early history of the goldfish is a murky topic. Several works helped me make sense of it: E. K. Balon's "About the oldest domesticates among fishes," *Journal of Fish Biology* 65 (2004); Svein A. Fosså's "Man-made fish: Domesticated fishes and their place in the aquatic trade and hobby," *OFI Journal* (February 2004); and Li Zhen's *Chinese Goldfish* (Morris Plains, NJ: Tetra Press, 1990).

Chapter Thirteen: Beard of the Crocodile

For more on the debate over specimen collection, see "Avoiding (Re) extinction," by Ben A. Minteer, James P. Collins, Karen E. Love, and Robert Puschendorf, and the responding letter "Specimen collection: An essential tool" by more than a hundred scientists, including Ralf Britz and E. O. Wilson, in *Science* 344 (2014). The data on coelacanth collection comes from Michael N. Bruton and Sheila E. Coutouvidis's "An inventory of all known specimens of the coelacanth *Latimeria chalumnae*, with comments on trends in the catches," *Environmental Biology of Fishes* 32 (September 1991).

The quotation from Charles Kingsley about the danger of pursuing new species appears in his book *Glaucus: Or, The Wonders of the Shore* (London: Macmillan and Company, 1855). The deep-sea port being built in Dawei is detailed in Thant Myint-U's *Where China Meets India: Burma and the New Crossroads of Asia* (New York: Farrar, Straus and Giroux, 2011). The FAO has concluded that 80 percent of the fifty thousand kilometers of major rivers in China are so degraded that they no longer support fish (http://www.fao.org/nr/water/aquastat/countries_regions/chn/index.stm).

PART V

INTO THE LAIR

Chapter Fourteen: The Paradox of Value

For more on John Audubon and Constantine Rafinesque, see Lynn Barber's *The Heyday of Natural History, 1820–1870* (Garden City, NY: Doubleday, 1980). The story of the fraudulent lungfish comes from Helfman et al. (see chapter 1).

The case of former Indonesian National Police chief detective Susno Duadji is complex, involving allegations of corrupt dealings in the wake of the bailout of Indonesia's Bank Century; a feud between the Indonesian National Police and the Indonesian Corruption Eradication Commission; and allegations that Duadji is being punished for testifying against a senior political official at a murder trial. Despite all this, Duadji was tried and convicted on the relatively minor offense of accepting a bribe related to a dispute with an arowana farm. I have relied on reporting by the BBC and the Economist Intelligence Unit to make sense of this long-running political drama.

For more on both the Lacey Act and the Fur Seal Act, see the third edition of Michael J. Bean and Melanie J. Rowland's *The Evolution of National Wildlife Law* (Wesport, CT: Praeger, 1997). John Lacey's "sucked orange" quote appears in his 1901 "Address to the League of American Sportsmen," New York. The issue of cattle DNA polluting the modern bison is described in "Out West, with the buffalo, roam some strands of undesirable DNA," the *New York Times* (January 9, 2007). For more on the passenger pigeon, see Joel Greenberg's *A Feathered River Across the Sky* (New York: Bloomsbury, 2014).

An authoritative history of CITES is available in Willem Wijnstekers's *Evolution of CITES*, 9th ed. (Châtelaine-Geneva, Switzerland: CITES Secretariat, 2011). I have also relied on the book *Endangered Species Threatened Convention: The Past, Present and Future of CITES* (London: Earthscan, 2000) edited by Jon Hutton and Barnabas Dickson. Of particular interest was Henriette Kievit's chapter, "Conservation of the Nile Crocodile: Has CITES Helped or Hindered?" Grahame Webb of Wildlife Management International, who chaired the IUCN Crocodile Specialist Group, provided much wisdom on crocodile ranching. For more on tiger farms, see, for example, Diane Toomey's interview with wildlife activist Judith Mills, "How Tiger Farming in China Threatens World's Wild Tigers," *Yale Environment 360* (January 20, 2015).

The quotation about a sea change in fish intelligence comes from "Learning in fishes: From three-second memory to culture" by Kevin N. Laland, Culum Brown, and Jens Krause, *Fish and Fisheries* 4 (2003). The studies I describe regarding how people value rarity in nature are "Fatal attraction: Rare species in the spotlight" by Elena Angulo, Anne-Laure Deves, Michel Saint Jalme, and Franck Courchamp, *Proceedings of the Royal Society B* 276 (2009); and "Rarity, value, and species extinction: The anthropogenic allee effect" by Franck Courchamp, Elena Angulo, Philippe Rivalan, Richard J. Hall, Laetitia Signoret, Leigh Bull, and Yves Meinard, *PLOS Biology* 4 (December 2006).

Chapter Fifteen: Fish Meets World

Tyson Robert's description of the batik arowana appears as "*Scleropages inscriptus*, a new fish species from Tanantharyi or Tenasserim River Basin, Malay Peninsula of Myanmar (Osteoglossidae; Osteoglossiformes)" in *Aqua* 18 (April 15, 2012). Several works helped me understand the "species problem," including Werner Kunz's *Do Species Exist?: Principles of Taxonomic Classification* (2012); Carl Zimmer's "What Is a Species?," *Scientific American* (2008); and a special issue on the subject in *Reviews in Fish Biology and Fisheries* 9 (1999). Stephen Jay Gould's conclusion that "there is surely no such thing as a fish" appeared in his essay "What, if anything, is a zebra?," *Natural History* (July 1981). For the risks that come with describing new species, see "Scientific description can imperil species" by Bryan L. Stuart, Anders G. J. Rhodin, L. Lee Grismer, and Troy Hansel, *Science* 26 (May 26, 2006).

Edward O. Wilson's autobiography, *Naturalist* (Washington, DC: Island Press, 1994), provides a firsthand account of the mid-twentieth-century "molecular wars." Nathan K. Lujan and Larry M. Page's op-ed "Libraries of Life," the *New York Times* (February 27, 2015), addresses the neglect of natural history collections. For a more detailed overview of the crisis, see *Collection Building in Ichthyology and Herpetology* edited by Theodore Pietsch and William D. Anderson Jr. (Lawrence, KS: American Society of Ichthyologists and Herpetologists, 1997).

In writing about the Amazon, I have relied on numerous sources, especially *The Smithsonian Atlas of the Amazon* by Michael Goulding, Ronaldo Barthem, and Efrem Ferreira (Washington, DC: Smithsonian Books, 2003). The report I mention on the Calderón is "Natural Resource Protection and Monitoring Commission to the Calderón River and Amacayacu National Park Buffer Zone," Fundación Entropika (March–November 2008).

Chapter Sixteen: Plan C

For more on the neotropical fish fauna, including the number of fish species in the Amazon, see *Check List of the Freshwater Fishes of South and Central America* by Roberto E. Reis, Sven O. Kullander, and Carl J. Ferraris (Porto Alegre, Brazil: EDIPUCRS, 2003). Michael Goulding's *The Fishes and the Forest: Exploration in Amazonian Natural History* (Berkeley: University of California Press, 1981) contains a detailed study of the silver arowana, which is far better understood than its Asian cousin. See, for example, "Potential threat of the international aquarium fish trade to silver arawana *Osteoglossum bicirrhosum* in the Peruvian Amazon," *Oryx* 40 (April 2006).

The US Bureau for International Narcotics and Law Enforcement Affairs held that the chemical glyphosate was "practically non-toxic to fish" in "Fact Sheet: Eradication of Illicit Crops: Frequently Asked Questions" (November 30, 2001). Since then, studies have challenged this claim. See, for example, "Toxicity and effects of a glyphosate-based herbicide on the Neotropical fish Prochilodus lineatus" by Vivian do Carmo Langiano and Cláudia Martinez, *Comparative Biochemistry and Physiology* 147 (March 2008). In March 2015, after the World Health Organization declared that glyphosate probably causes cancer in humans, Colombia halted the aerial spraying of crops used to make cocaine. Coca farmers are now reportedly transitioning to farming arowana as a more profitable undertaking ("Trading coca for fish farms," the *Toronto Star* [August 24, 2013]).

On the plane to Leticia, I was reading Ross Socolof's *Confessions of a Tropical Fish Hobbyist* (Bradenton, Florida: Socolof Industries, 1996), an insider's account of the twentieth-century aquarium trade. The study on the depletion of the arapaima population is "Understanding fishing-induced extinctions in the Amazon" by Leandro Castello, Caroline Chaves Arantes, David Gibbs Mcgrath, Donald James Stewart, and Fabio Sarmento De Sousa, *Aquatic Conservation: Marine and Freshwater Ecosystems 25* (October 2015). That more of the Amazon rain forest has been destroyed in the last fifty years than the previous five centuries comes from Scott Wallace's *The Unconquered: In Search of the Amazon's Last Uncontacted Tribes* (New York: Crown Publishers, 2011).

Chapter Seventeen: Here Be Dragons

For a fantastic account of Theodore Roosevelt's expedition through the Amazon, see Candice Millard's *The River of Doubt: Theodore Roosevelt's Darkest Journey* (New York: Doubleday, 2005). Richard Conniff's "Shocking truth about piranhas revealed!," the *New York Times* (January 3, 2013), provides a counterpoint to Roosevelt's uncharitable characterization of the piranha. For more on the candiru, see Stephen Spotte's *Candiru: Life and Legend of the Bloodsucking Catfishes* (Berkeley, CA: Creative Arts Book Company, 2002).

James Serpell's theory that domestication began with pet-keeping appears in his book *In the Company of Animals: A Study of Human-Animal Relationships* (Oxford: B. Blackwell, 1986). The contending theory of self-domestication was explored at a 2014 conference on Domestication and Human Evolution hosted by the Salk Institute, which can be viewed online at http://carta.anthropogeny.org/events/domestication-and-human-evolution.

Alfred Russel Wallace's *Fishes of the Rio Negro* (São Paulo: Editora da Universidade, 2002) was finally published 150 years after Wallace managed to rescue the manuscript from the flames of his sinking ship.

Epilogue

The story of the Arrow People and the Javari Valley is recounted in Scott Wallace's riveting book *The Unconquered: In Search of the Am-*

azon's Last Uncontacted Tribes (New York: Crown Publishers, 2011). The estimate of the amount of fish collected from the wild as pets versus food appears in John Dawes, "International Experience in Ornamental Marine Species Management—Part I: Perspectives," *OFI Journal* 26 (February 1999). I first read about the study in Gene S. Helfman's *Fish Conservation: A Guide to Understanding and Restoring Global Aquatic Biodiversity and Fishery Resources* (Washington, DC: Island Press, 2007), which provides a comprehensive overview of the myriad issues affecting freshwater fish, including the aquarium trade and the problem of invasive species. For more on dam construction threatening the Amazon, see "Proliferation of hydroelectric dams in the Andean Amazon and implications for Andes-Amazon connectivity," by Matt Finer and Clinton N. Jenkins, *PLOS One* (April 18, 2012).

I am indebted to Sven Kullander for tallying the number of fish named after a penis on his delightful blog *Fish Matter*.